D0122737

Journey of the Universe

Journey
of the
Universe

BRIAN THOMAS SWIMME

AND

MARY EVELYN TUCKER

Yale UNIVERSITY PRESS

NEW HAVEN & LONDON

Yale University Press books may be purchased in quantity for educational, business, or
promotional use. For information, please e-mail sales.press@yale.edu (U.S. office)
or sales@yaleup.co.uk (U.K. office).

Designed by Mary Valencia

Set in Minion type by
Keystone Typesetting, Inc.

Printed in the United States of America by Courier.

Library of Congress Cataloging-in-Publication Data
Swimme, Brian.
Journey of the universe / Brian Thomas Swimme, Mary Evelyn Tucker.
p. cm.
Includes bibliographical references and index.
ISBN 978-0-300-17190-7 (hardback)
1. Cosmology—Popular works. 2. Astronomy—Popular works. 3. Evolution—Popular works.
I. Tucker, Mary Evelyn. II. Title.
QB982.S95 2011
523.1—dc22
2010051084

A catalogue record for this book is available from the British Library.

This paper meets the requirements of ANSI/NISO Z39.48-1992 (Permanence of Paper).

10   9   8   7   6   5   4

To Nancy Klavans,
friend and fellow traveler on the journey.
With fond memories of our times together,
especially drinking tea at the Japanese garden
in Golden Gate Park.

CONTENTS

ACKNOWLEDGMENTS

A WORK SUCH AS THIS HAS BEEN A GREAT JOURNEY, and many people have shared that journey with us. We are immensely grateful to them and to all those who have inspired us along the way to create a book and a film on the Journey of the Universe. Thomas Berry was with us from the outset, giving us a sense of the "great work" of our time.

This great work is shared by so many of our supporters including Nancy Klavans, Marty and Wendy Kaplan, Bruce Bochte, Lavinia Currier, Susan O'Connor, Diana Blank, Diane Ives, David Orr, Nancy Schaub, Jean Berry, Bokara Legendre, Peter Teague, Barbara Sargent, Barbara Cushing, Richard Rathbun, Albert Neilsen, Clare Hallward, Roger Cooke and Joan Cirillo, Edith Eddy, and, most especially, Mary Elizabeth Tucker and Jeanne Swimme.

In addition, various foundations have assisted our work on the *Journey* book and the film including Germeshausen Foundation, Kendeda Sustainability Fund, Compton Foundation, Englehard Foundation, Foundation for Global Community, Kalliopeia Foundation, Lewis Foundation, New Priorities Foundation, Nathan Cummings Foundation, Sacharuna Foundation, Tara Foundation, and the Tides Foundation.

To thank the readers of this manuscript is a special joy. So many of you have been such significant collaborators over the years. We are especially grateful to scientists Craig Kochel, Larry Edwards, and Terry Deacon as well as Marc Bekoff, Barb Smuts, Ann Berry Somers, Scott Sampson, Todd Duncan, Russ Genet, Michael Wysession, and Claude Bernard. Ursula Goodenough's perceptive insights and careful readings of the manuscript were immeasurable gifts.

A deep bow of gratitude to our colleagues in the humanities for their comments: John Grim, Steven Rockefeller, Brian Brown, Miriam MacGillis, David Kennard, Anne Roberts, Rick Clugston, Marty Kaplan, Heather Eaton, Anne Marie Dalton, Chris Chapple, Margaret Brennan, Louis Herman, Neal Rogin, Kym Farmer, John Cobb, Catherine Keller, Larry Rasmussen, and John Haught. Heartfelt appreciation for conversations with colleagues at the California Institute of Integral Studies, especially Robert McDermott, Rick Tarnas, Sean Kelly, Elizabeth Allison, Eric Weiss, Jacob Sherman, and Aaron Weiss, all of whom gave important feedback during a memorable afternoon symposium.

Thanks to the extraordinary nature writers Kathleen Dean Moore, Scott Russell Sanders, and Alison Hawthorne Deming for their astute reading. Gus Speth appreciated this enterprise in ways we could not have foreseen. Gratitude abounds!

Special mention must be made of the invaluable assistance of Arthur Fabel with the bibliography. His extensive knowledge of this literature has been honed for over thirty years. Cynthia Brown also made helpful suggestions. Our webmasters Elizabeth McAnally and Sam Mickey provided a brief annotated bibliography for our website.

At Yale the Dean of the School of Forestry and Environmental Studies, Sir Peter Crane, has been unflagging in his support. At Yale University Press we have benefited from the careful editing of Jeff Schier. We are also indebted to the reader's reports from Tom Lovejoy, George Fisher, David Orr, and J. Baird Callicott. Without Tara Trapani's painstaking preparation of the manuscript we would never have met our deadline. She has been remarkable.

Special thanks are reserved for Yale University Press's science editor, Jean Thomson Black. Her skillful attention to every detail of the process is indicative of why she has overseen so many remarkable science books at Yale for some twenty years. Jean understood this book from the outset and shared our admiration for the writings of Loren Eiseley, who skillfully blended the sciences and humanities.

Finally it is to our partners, John Grim and Denise Swimme, that we extend heartfelt gratitude. Your smiles are what sustain the journey.

Journey of the Universe

ONE

Beginning of the Universe

Imagine experiencing Earth's beauty for the first time—its birds, fish, mountains, and waterfalls. Imagine, too, the vastness of Earth's home, the universe, with its numerous galaxies, stars, and planets. Surrounded by such magnificence, we can ask ourselves a simple question: Can we find a way to sink deeply into these immensities? And if we can, will this enable humans to participate in the flourishing of life?

This book is an invitation to a journey into grandeur—a journey into grandeur that no previous generation could have fully imagined.

We are the first generation to learn the comprehensive scientific dimensions of the universe story. We know that the observable universe emerged 13.7 billion years ago, and we now live on a planet orbiting our Sun, one of the trillions of stars in one of the billions of galaxies in an unfolding universe that is profoundly creative and interconnected. With our empirical observations expanded by modern science, we are now realizing that our universe is a single immense energy event that began as a tiny speck that has unfolded over time to become galaxies and stars, palms and pelicans, the music of Bach, and each of us alive today. The great discovery of contemporary science is that the universe is not simply a place, but a story—a story in which we are immersed, to which we belong, and out of which we arose.

This story has the power to awaken us more deeply to who we are. For just as the Milky Way is the universe in the form of a galaxy, and an orchid is the universe in the form of a flower, we are the universe in the form of a human. And every time we are drawn to look up into the night sky and reflect on the awesome beauty of the universe, we are actually the universe reflecting on itself.

And this changes everything.

STORY

Every culture organizes itself with its central stories, in both written and oral forms. Such stories contain that which each culture holds as most valuable, most useful, most essential, and most

beautiful. They are regarded as containing compelling orientations toward the most enduring human challenges. Some of these
stories have been so deeply valued they have been told over and
over by many generations. Homer's *Odyssey*, for example, has
been passed down in the West for perhaps twenty-eight centuries.
Or, in south Asia, the stories of the *Mahabharata* have been told
for well over two millennia. In distinct and invaluable ways, these
stories and many others continue to shape billions of humans
around the planet.

While such stories will no doubt be told far into the future, a
new integrating story has emerged. Even though it is only a few
centuries old, it has already begun to change humanity in crucial
ways. This is the story of the universe's development through
time, the narrative of the evolutionary processes of our observable
universe. This story has, and will continue to have, many different
names. But if we can think of the New Testament as that which
tells a Christian story, and of the *Mahabharata* as that which tells a
Hindu story, perhaps the simplest description of this new narrative is that it tells a universe story.

One of the differences between a universe story and more
traditional narratives is that with this newer story we have a "story
of the story"—a historical account of how our awareness of this
universe story came forth. This began in the sixteenth and seventeenth centuries, when we realized that Earth was not stationary
but was moving around the Sun. In the eighteenth century this idea

was extended when we came to realize that the human mind was not static, nor was human society; instead, both had forms and structures that had emerged over many centuries. Then, in the nineteenth century, we discovered that the life forms themselves had gone through significant transformations over time. Even the rocks were not inert but were also in the process of deep change throughout geological time. Finally, in the twentieth century, we came to see that the stars too had changed dramatically, as had the galaxies, and, most astonishing of all, the entire observable universe had passed through a series of irreversible transformations.

This immense journey evokes wonder from scientists and nonscientists alike. And it challenges some religious traditions to rethink or expand their worldviews. Certainly Copernicus was aware of the radical nature of his discovery of a heliocentric solar system and hesitated to reveal his work. Darwin also agonized over the revolutionary implications of his views regarding life's emergence. We are still struggling with the changes of worldviews that Copernicus and Darwin and many other scientists have presented to us over the past five centuries. And why? Because this is such a comprehensive story that it challenges our understanding of who we are and what our role might be in the universe. Are we here by chance, by necessity, by serendipity, or on purpose? What is the nature of creativity in this changing universe?

It will take time to answer these questions more fully and to integrate this universe story into our diverse human cultures around the world. *Journey of the Universe* is intended not to override or ignore these other stories, but rather to bring into focus the challenge of creating a shared future. The great opportunity before us today is to tell this new universe story in a way that will serve to orient humans with respect to our pressing questions: Where did we come from? Why are we here? How should we live together? How can the Earth community flourish?

BIRTH OF THE UNIVERSE

Let's begin at the very beginning. How did it all start?

An awesome question, certainly, but it appears there really was a beginning. Some scientists refer to this as the Big Bang. Let's think of it as a great flaring forth of light and matter, both the luminous matter that would eventually become stars and galaxies and the dark matter that no one has ever seen. All of space and time and mass and energy began as a single point that was trillions of degrees hot and that instantly rushed apart.

The discovery that the universe has expanded and is still expanding is one of the greatest of human history. In the modern West, the common understanding had been that the universe is simply a vast space in which things existed—large things like stars and small things like atoms. Scientists knew that matter changed form in the universe, but they assumed that the universe as a whole was not changing. That assumption proved to be mistaken, for the universe is unfolding and has a story—a beginning, a middle (where we are now), and, perhaps in some unimaginable future, an end.

One of the scientists responsible for this great discovery is Edwin Hubble. In the 1920s, atop Mount Wilson in southern California, he focused the hundred-inch telescope on the night sky. He was trying to determine if our Milky Way was the only galaxy in the universe. Not only did he discover that the universe is filled with galaxies; he also determined that all of them were rushing away from each other. Building on Hubble's work, scientists have learned that the entire observable universe was once smaller than a grain of sand, a tiny dot that began with a massive inflation that has been carrying matter apart for billions of years. The universe arose with a titanic expansion.

But there is another fundamental force at play in our universe: a force of attraction, pulling things together—a force we call gravity. The universe expanded and cooled, and gravity pulled some of the matter together to form the galaxies and stars. These two opposing dynamics, expansion and contraction, were the dominant powers operating at the beginning of the universe. The expanding universe was causing matter to move apart from the tiny seed point of its beginning. Gravity was drawing some of this matter back together again. We now know that the universe as a whole, from the beginning, has been shaped by these two opposing and creative dynamics.

This double process is wonderfully reminiscent of life, of the movement of breath and of blood. Our lungs expand and contract. Our heart expands and contracts. Within such primordial movement we come into existence. In a very literal sense, our lives are possible because of this in-and-out rhythm of the universe. As we fill our lungs with breath are we mirroring the large-scale dynamics of our universe? At the very least we can say that because of the great exhalation of the universe, life and humanity have emerged and are breathing within it now.

NUCLEI AND BONDING

In the beginning, the universe brought forth quarks and leptons, the elementary particles, and within a few microseconds the quarks combined to form protons and neutrons that churned ceaselessly in a thick and gluey form of matter called plasma. There was almost no structure in the universe. These quanta would collide, would interact with one another, and then would scatter apart to collide with different partners millions of times each instant.

Our current mathematical model of the early universe asserts that even in the first few minutes, more structures began to emerge. The elementary particles began forming stable relationships. A single neutron might interact with a single proton, and instead of scattering away they would remain bonded together. At first these new bonds were quickly torn apart by other particles. But as the universe continued to expand and cool, these primordial couples and triads began to survive.

Amidst such bonding and dissolution, the universe moves toward increasingly complex communities. These simple nuclei were the very first of the complex communities among the elementary particles. Intriguingly, all relationships carry a cost, even at this quantum level. A neutron does not simply adhere to a proton. Rather, both the neutron and the proton have to undergo a transformation for the bonding to occur. The proton and the neutron each give over part of their mass, which becomes a flash of light released into the universe. Who could have imagined this? Who could have guessed that the creation of a quantum community would require the contribution of the mass of the particles? Or that its creation would be accompanied by a flash of light?

Even from the first moments, our universe moved toward creating relationships. Certainly, in a theoretical sense, we can imagine that things could have been different. We can theorize about a different kind of universe, a universe that would have taken the form of disconnected particles, a universe that would never have formed bonded relationships. Such a universe would consist of trillions upon trillions of these tiny particles, each one completely independent of the others. But in our observable universe, various forms of bonding are inescapable. Even moments after the birth of the universe, the simple nuclei were brought forth, an act that required vast amounts of mass throughout the universe to be transformed into light. The entire universe was permeated with a new burst of radiation as protons and neutrons fused together into the first nuclei. This bonding is at the heart of matter.

TIMING AND CREATIVITY

Within an unimaginably vast and complex universe, we seek meaningful orientations in order to live an integral human life. Humans have always sought answers to questions such as, What is the nature of the universe? What is our role? By pondering such questions we are hoping to become more fully and deeply alive in this emerging planetary era in which we find ourselves.

The fundamental images we hold of the universe are central to the whole process of exploring meaning. An image cannot carry the fullness of the universe, and thus we need multiple images or metaphors. Already we have considered at least three images of the universe. We spoke of the universe as unfolding its structures from a tiny ball. We spoke of the universe as lungs breathing and as a heart expanding and contracting. And we spoke of the evolution of matter with the implicit suggestion that the universe is filled with complexifying communities.

Another image presents itself when we consider the origin of these nuclei soon after the birth of the universe, the image of a developing seed. When a seed germinates, it will initially focus mainly on bringing forth roots; later it will concentrate on constructing leaves. The process of its growth is a complex and creative orchestration. Similarly, the universe in its first moments focuses on building nuclei. The process continues for a brief time and then stops, and other processes emerge. The astonishing fact is that if the universe had continued building nuclei all the way up to iron, for example, iron nuclei would have predominated for all time.

But the universe was expanding and cooling, and as quickly as the conditions for building nuclei emerged, they changed. After that brief moment when all of the light nuclei had been created, there was a shift. Something new was about to emerge, in a way analogous to the development of a plant from a seed. This dynamic of timing will appear again and again over the fourteen billion years of cosmic unfolding.

EXPANSION AND EMERGENCE

One of the most spectacular features of the observable universe is the elegance of its expansion. If the rate of expansion had been slower, even slightly slower, even one millionth of a percent slower, the universe would have recollapsed. It would have imploded upon itself, and that would have been the end of the story.

Conversely, if the universe had expanded a little more quickly, even one millionth of one percent more quickly, the universe would have expanded too quickly for structures to form. It would have simply diffused into dust, with no structures to bring forth life.

What we've discovered is that we are living in a universe that is expanding at just that rate necessary for life to emerge. When scientists first discovered this they were filled with a desire to understand this amazing fact. What happened in the past to make our universe like this?

As mathematical cosmologists began to probe this mystery of what gave rise to a life-generating universe, they came up with a theory, first articulated by Alexei Starobinsky at the Landau Institute at the Russian Academy of Sciences and later given a more comprehensive form by Alan Guth now at MIT. Drawing on the ideas of Albert Einstein and his General Theory of Relativity, these cosmologists realized that at the beginning of time gravity exerted a form of repulsion rather than attraction. It was precisely this repulsive form of gravity that forced the universe to expand right to the critical expansion rate. In other words, the universe utilized its own inflationary mechanism to expand rapidly to the rate that enabled it to bring forth structure and life.

When the celebrated physicist Freeman Dyson was reflecting on all of this and trying to make sense of it, he realized that he had come to feel at home in the universe in a new way: "The more I examine the universe and study the details of its architecture," he wrote, "the more evidence I find that the universe in some sense must have known that we were coming."[1] Of course, humans were not present in any explicit sense at the beginning, but Dyson is suggesting that we are now learning ways in which life was implicitly present in the very dynamics themselves, from the very first moment.

ATOMS AND ATTRACTION

Attraction is at the heart of creativity at all levels of being. When the universe was less than half a million years old the plasma was a dense, thick, gluey form of matter whose components were primarily helium nuclei, hydrogen nuclei, and electrons. All of this was permeated with an ocean of light. But as the universe continued to expand and cool there came a moment of transformation when the electrons and protons came together to form the first atoms.

The structure of atoms is governed by the electromagnetic interaction between electrically charged particles. Oppositely charged particles pull each other together. This electrical attraction drew electrons (negatively charged) and protons (positively charged) to form hydrogen and helium atoms. Hence, the universe as a whole transformed itself from one vast plasmic ocean of elementary particles into endlessly billowing clouds of much larger atoms.

We cannot fully explain why a proton is attracted to an electron. Saying that opposite electrical charges attract one another does not address the mystery of why this is so. Nothing outside is pushing them together. They are not being forced together by something called "electromagnetic interaction." Rather, it is by their very nature that they are drawn to each other.

We are left marveling over the fact that the allurement between opposites gave birth to the atoms. And who is it that is marveling over this fact? It is none other than we humans—a much later development of these very atoms. The attraction between a proton and an electron is not just another disconnected fact about our universe. Attraction between a proton and an electron is a way in which the universe gives rise to ever greater complexity, which, after some fourteen billion years, includes us.

THE UNIVERSE BECOMES TRANSPARENT

Scientists made a fascinating discovery that a change at the micro level, such as the birth of atoms, can actually alter the overall qualities of the macro universe. We can begin to appreciate this dynamic by reflecting further on what took place with the appearance of the first atoms—the universe became transparent.

This transformation can be compared to a fog lifting. In a fog we cannot see anything in the distance because the light is scattered by the water droplets of the fog. It was the same with the plasma in the early universe. Particles of light could travel only a fraction of an inch before they were absorbed and then scattered by an electron or a proton.

But when the electrons and protons began bonding together into electrically neutral atoms, a light particle was no longer scattered as it encountered yet another electrically charged particle. Light could suddenly travel in straight lines. Some of this light might disappear should it be absorbed, for instance, into the matter of a cold cloud of gas. But much of this primordial light continued unimpeded on its journey for billions of years. During this time the universe entered further into its deep, complexifying processes. Thus, today, when we train our sensitive instruments on the night sky we are able to detect these photons from the beginning of time and learn their story of the nature of things near the very birth of the universe.

The appearance of atoms enabled the universe to enter an entirely new phase of its creativity. If no atoms had formed, the luminous matter would continue in the form of plasma, which would be clumped in various ways by the dominant presence of dark matter. The flashing and scintillating light from the various interactions among the leptons and hadrons would continue for billions of years. But with the appearance of atoms, new possibilities arose. The universe could now bring forth entirely new structures—the stars and galaxies.

Thus an event taking place on the micro scale—the formation

of the atoms of hydrogen and helium—affected the overall story of the macrocosm. There's something amazing about a universe whose overall journey depends, in critical moments, upon the transformations taking place in the microcosm. We can begin to contemplate an idea that is remarkable: perhaps the nature of the universe as a whole is shaped by the creativity of its parts.

TWO

Galaxies Forming

How are we to understand the beauty of the universe? We are surrounded by beauty. What brought it into being? Where does the intricacy of a dragonfly or a lilac come from?

Let's consider the birth and development of galaxies. Even a century ago we knew only about one galaxy in the entire universe: our own Milky Way. Over the course of the twentieth century we discovered nearly a hundred billion galaxies. Each of these contains several billion stars. What does this mean for understanding our place amidst such vastness?

We are only now entering into an ongoing reflection regarding the origin of galaxies. Scientists have made several crucial discoveries. When the universe was almost half a million years old, it was like a vast cumulus cloud billowing out. We can imagine a scenario where this cloud, composed of both luminous and dark matter, just keeps expanding forever, but in the actual universe this cloud instead broke into numerous, smaller clouds. Each of these clouds pulled itself out of the cosmic expansion of the universe and collapsed into a single galaxy or a cluster of galaxies. Thus each jelled and remained the same size while the distances between the clouds continued to increase. As a result, each cloud could start on its own unique journey.

We can see here something of the nature of creativity in the universe. To enter its own creative development, a dynamic system will sometimes pull itself away from its larger enveloping network. As long as a system is tightly held within a larger system, it is dominated. But as it becomes free its intrinsic potentialities come forth and are amplified so that something new can enter into existence.

A further insight into the creativity of the universe follows from this question: What caused the fracture of the initial cloud into all these smaller clouds? For the power that breaks up this cloud is the power that sets the universe in a new direction. This power is responsible, in a primordial sense, for the advent of the galaxies.

Scientists have discovered that a series of waves passing through the universe was responsible for fragmenting the initial cloud. And the origin of these waves? This is the biggest surprise. These waves had their origin in the birth of the universe itself. In the initial flaring forth the universe was pervaded with waves. These waves, which are fluctuations in the density of matter, grew as the universe expanded. Eventually they broke the universe apart so that galaxies might form.[1]

We know now that the galaxies emerged from the primordial vibrations in the birth of the universe. These vibrations in matter certainly had a special power of creativity. Perhaps we can regard them as a kind of music, a "music of the spheres."

Pythagoras, who laid the foundations for mathematical science twenty-six centuries ago, would certainly be delighted, for his intellectual heirs have discovered that the billions of galaxies were formed by a cosmic music that moved the universe into the next phase of its journey.

GALAXY CLUSTERS AND A MULTICENTRIC UNIVERSE

How can we orient ourselves within this cosmic music, amidst the vast structures of the universe?

Each culture has had its own particular understanding of the universe, enabling its members to orient themselves with respect to space and time. One of the most fundamental orientations for humans concerns the center of things. Again and again, we have asked ourselves, where is the heart of the universe?

Each culture has its own answer regarding the center. Some locate it on a special mountain such as Mount Kailasa in Tibet or Mount Kilimanjaro in Africa. Others designate a particular city, such as Jerusalem, Rome, or Mecca in the West, and Beijing, Varanasi, or Jogjakarta in Asia. Such cities become places of religious pilgrimage or seats of political power.

We can easily appreciate how significant such cities are for humans. To be related to the center is to receive a special value. For instance, citizens of the city at the center of the world enjoy a status not readily extended to someone from the periphery. And certainly any laws or decrees issuing from the center carry a special authority.

The five-hundred-year enterprise of modern Western science has also been concerned with identifying the center of the universe, and this effort has led to a series of "de-centerings." We have learned that our former ideas concerning the center were not the full story. Perhaps the most famous contribution to de-centering the human world was when we discovered that the Earth was not the unmoving center of things, but was rather in motion around the Sun. This was first conjectured by Aristarchus in the third century BCE on the island of Samos in Greece, and later was independently discovered in Europe by Copernicus in 1543. Within a few centuries, our ongoing investigation led to the realization that although the Sun, indeed, was the center of the solar system, it was not the center of the universe. In 1918 Harlow Shapley provided evidence indicating that the Sun was moving in a great ellipse around the center of the Milky Way galaxy. This de-centering process was carried still further when Edwin Hubble and others, in the 1920s, discovered that the Milky Way was not the central galaxy of the universe. Rather, our Milky Way is just one galaxy in a universe filled with galaxies.

When scientists discovered that the observable universe contains a hundred billion galaxies, they were stunned. For scientists and nonscientists alike, absorbing the significance of living within such a vast, evolving universe is an ongoing challenge.

A surprising development in the second half of the twentieth century has led to an entirely new understanding of center. This understanding goes against common sense and is a challenge to absorb fully. For what we have come to realize is that there is not one center, but millions. Each supercluster of galaxies is at the very center of the expansion of the universe. We live in a multicentered universe and are only now awakening to this new discovery.

For instance, our Milky Way galaxy is one of a several dozen galaxies revolving around each other. This system as a whole is moving around the Virgo Cluster of galaxies. There are also other groups revolving about the Virgo Cluster, and this entire system is called the Virgo Supercluster. We can picture this as something like planets swirling about a central star, where the planets are the individual galaxy clusters and the central star is the massive Virgo Cluster. What we have learned is that this Virgo Supercluster is at the very center of the cosmic expansion.

What is striking and counterintuitive is that the other superclusters throughout the universe are also at the center of the cosmic expansion. To visualize this, picture the universe as a loaf of raisin bread rising, where each raisin is a supercluster of galaxies. As the loaf grows larger, and we imagine ourselves on one of the raisins, we would see all the other raisins moving away from us. We would also conclude that we were not moving because we would not be moving through the bread. It would not matter which raisin we chose. Such is the nature of the large-scale universe. In terms of the expansion, each supercluster is stationary, while all the other superclusters are expanding away from it.

This staggering new perspective is causing a massive shift in our understanding of how we imagine our own place, our home. We realize now that we dwell in one center in a universe that is composed of millions of such centers. While this is difficult to comprehend, we are learning, nonetheless, to orient ourselves with wonder and awe in the midst of these immensities.

SPIRAL GALAXIES AND THE BIRTH OF STARS

What is the nature of our center? Is it a good place? A safe place? Such questions press into our awareness no matter where we live on Earth. But what if we ask the questions not of our neighborhoods, or of our nation or our planet, but rather of the Milky Way galaxy?

The most powerful feature of our galaxy is its spiral structure. When scientists first detected the spiral arms they concluded that they were formed of matter and that they were spinning about the center of the galaxy. But this proved to be a mistaken theory. By correcting their mistake scientists learned one of the most astonishing features of creativity in the universe.

An arm of the Milky Way is not a static structure. It is rather an effect of huge gravitational waves, called density waves, that are pulsing through the Milky Way. In every spiral galaxy, the density waves cause the collapse of gas clouds into massive stars that burn brilliantly for a million years and then explode or die out. As this happens the wave passes farther on and ignites the formation of a new set of stars, giving the overall impression of something like a spinning pinwheel.

This spiral structure of a galaxy enables it to continue creating stars. It is in this sense always new, always capable of fresh, creative action. Thus, by virtue of their architecture, spiral galaxies are the birthing galaxies in the universe.

Elliptical galaxies, which have roughly the shape of an egg, do not have this creative capacity. Most of the stars that exist in an elliptical galaxy are doomed to die out without being replaced. Ellipticals lack the architectural form necessary for creating new stars.

The fascinating discovery is that the creativity of the universe is not evenly distributed but is concentrated in particular places. At the level of galaxies, creativity is concentrated in the spirals. But within a spiral galaxy there are particular places where creativity is more intense than in other places. And within these places there are particular regions where the intensity reaches its maximum. To find oneself in the midst of a nested domain of creativity is to move into the depths of creativity itself. To be outside the locus of creativity would be a kind of exile.

We awaken to existence and discover ourselves in the inner circles of creativity. Held by the embrace of a spiral galaxy, we enter into a multilayered and seemingly infinite fecundity.

GALACTIC RELATIONSHIP AND MUTUAL EVOCATION

The galaxies themselves come forth amidst immense creativity. The dynamics of the universe ignite creativity in new forms whenever possible. This occurs through processes that can be described as mutual evocation.

One galaxy that makes this drama clear is a satellite to the Milky Way called the Large Magellanic Cloud, or LMC for short. Though our knowledge of its history is far from settled, some astronomers speculate that LMC began as a spiral galaxy, but some cataclysm took place billions of years ago and its spiral structure was destroyed. Perhaps this occurred in a head-on collision with another galaxy. Or perhaps LMC passed by a larger galaxy whose gravitational attraction was too much to bear and tore its large-scale structure apart. Whatever the trauma was, it led to a collapse of its ability to create stars. Thus was LMC stripped of the promise it had when it came forth as a young galaxy. LMC was abandoned. It drifted about, each star's death just another step toward the final darkness that now awaited it.

But then something happened. After billions of years, LMC was drawn into a gravitational relationship with our Milky Way galaxy. LMC began a new orbit that would lead to a new destiny. In a gravitational relationship each member is changed by the interaction. The gravitational tidal force issuing from the Milky Way penetrated into the system of stars that formed LMC, and the structure of this smaller galaxy began to change. A regeneration of LMC was occurring in the presence of the Milky Way.

And then an awakening occurred. A burst of star-making activity appeared in one of the dormant regions of LMC. For billions of years LMC had drifted about, barren and dying. Now, suddenly, its potentiality was ignited through this interaction and new stars were evoked into being in all their brilliance.

THREE

The Emanating Brilliance of Stars

Why are we so fascinated by the stars? Some of our ancestors thought stars were gods. Still others thought the stars were angels pouring forth virtue upon the Earth. Contemporary scientists refer to stars as giant balls of gas.

The need to orient ourselves with respect to the stars continues, but the way that twenty-first century humans approach this challenge includes a growing base of knowledge about the stars that previous generations did not enjoy. Perhaps the most significant discovery is that stars are self-organizing processes. They are not just unchanging bright objects in the night sky. Stars proceed through stages of development that enable their radiance to come forth.

What is the ultimate origin of a star's radiance? It comes from the intense compression of matter under the force of gravity. But what is the origin of this gravity? Strictly speaking, gravity is an effect of mass. Consider a vast cloud of hydrogen and helium that is destined to collapse into a future star. The gravitational attraction that causes the cloud to implode is generated by the mass of the cloud itself. In other words, the mass of the future star creates the gravity necessary to give birth to the star itself. In that sense, each star is a self-generating event.

And stars not only shine. They resonate, they communicate. Humans throughout history on every continent and in every culture have been stunned by the presence of stars in the vastness of the night sky. They have meditated on the beauty of the Big Dipper. So deeply moved by the majesty of the constellations and by the ineffable majesty emanating from the brilliance of stars, many have built their lives around them. They have imagined ways of not only organizing their personal lives but even patterning civilizations around the beauty and order found there.

In many cultures throughout history humans intuited that they descended from the stars, even before they had the empirical evidence from science that our bodies were formed by the elements forged by the stars. Humans felt something in the depths of the night as they contemplated the presence of the stars. They began to suspect that the meaning of their lives went far beyond what preoccupied them during the urgencies of the daytime world. They knew in their hearts that their journey and the radiance of the stars were interwoven.

THE BIRTH OF STARS

The essence of the universe story is this: the stars are our ancestors. Out of them, everything comes forth. The stars are dynamic entities. They have a birth. They go through a development. They come to an end, sometimes a dramatic end. Here's their story.

The birth of a star begins with a cloud of hydrogen and helium imploding under the influence of gravity. The cloud shrinks moment by moment. As the atoms draw themselves together into ever-tighter spaces, they collide and vibrate with energy. After each collision they gradually heat up. Even a cloud that starts out at temperatures hundreds of degrees below zero will slowly become warmer as the eons pass.

During this increase in temperature the process of star birth recapitulates processes that were active at the time of the origin of the universe. As the clouds of hydrogen and helium heat up to several thousand degrees, the atoms begin to melt down. The hydrogen atoms dissolve back into being protons and electrons, which then move about in the core of the protostar as freely interacting elementary particles.

The culminating moment, the very birth of the star, takes place when the temperature reaches ten million degrees. When the elementary particles get this hot they fuse into new stable relationships. This is similar to what took place in the early moments of the universe when the first nuclei were formed. The star thus has the capacity to activate creative processes that were at work billions of years ago. Such originating creativity is woven through space and time, waiting to be ignited. Humans in every culture have invented myriad ways in which this primal creativity could be accessed for the collective human journey.

THE ACTIVITY OF STARS

For stars, creativity depends on maintaining a state of disequilibrium with respect to surrounding space. It is the dynamic tension between gravity and fusion that enables the star to maintain this seething disequilibrium.

The power of gravitational attraction within a star presses toward total collapse of the star. The power of nuclear fusion, where protons and neutrons fuse together and release energy in the center of the star, aims at expansion: matter is literally pushed outward, the opposite of collapse. If either of these powers comes to dominate, the star's life ends. The star exists only because these two powers are kept in creative tension for billions of years.

The atoms in a star have a fundamental resistance to being

crushed because the electrons of one repel the electrons of the other. If the gravitational attraction is strong enough, the heat that results from this resistance causes atoms to dissociate into free electrons and nuclei.

But gravity does not stop here. The entire process is repeated at the level of the nuclei. Nuclei repel each other, but if the overall gravitational "crushing" is powerful enough, this resistance can also be overcome. The protons and neutrons in adjacent nuclei are brought so close to each other that they can fuse into the stable configuration of a new nucleus. This fusion process converts hydrogen nuclei into helium nuclei, thus releasing enough energy to push the star outward and stave off further collapse.

The star then exists in between extremes. On one side there is gravitational collapse; on the other is thermonuclear fusion and outward pressure. Thus the star exists not in a world of stasis but in a realm of seething disequilibrium. Because the star holds itself in this far-from-equilibrium realm, it is capable of creating helium nuclei out of elementary particles.

This is one of the most amazing discoveries in the history of science. Stars are fiery cauldrons of transformation. Stars are wombs of immense creativity. And one can wonder if these complex interactions that we see in the stars reflect deep patterns of creativity in other domains of the universe. Certainly there are similarities in the human world. Beset with strong emotions of attraction and repulsion, we can, even so, develop emotionally charged bonds that become the foundation for decades of creative action.

There is deep ambiguity threaded throughout that may result not simply in communion but also in collapse. But isn't this also the nature of the universe—both dangerous and inviting? How do we discover ourselves in forces that are simultaneously fearful and attractive? How do we live amidst shimmering disequilibrium? One thing seems certain: the universe, navigating between extremes, presses ever further into creative intensities.

THE EXPLOSION OF STARS

One of the greatest costs of creativity in the universe is the supernova, an exploding star. The astonishing fact is that many large stars are destined to explode. The energy expended in this event is unrivaled by anything else in the universe. The power of a supernova is equivalent to that of an entire galaxy with a hundred billion shining stars.

Stars do everything possible to avoid such an end. For a star twenty times the size of our Sun, the first challenge comes only ten million years after its birth. Throughout those first ten million years, the star has maintained its state of seething disequilibrium by fusing hydrogen nuclei into helium nuclei in its core. But eventually there is no hydrogen left in the core to fuse. It has all been transformed into helium nuclei. So the outward-pushing energy that came from the fusion processes stops.

When this happens, gravity causes the star to collapse into a smaller space. Without any fusion-derived energy pushing out, the star can press itself ever smaller, but as it does so the core of the star heats up until it reaches the temperature necessary to fuse helium into carbon. Now, once again, the star can settle down into a semistable state, for the new blast of energy in its center is enough to hold back the huge force of gravity. This stable state will continue so long as there is helium to fuse. But once the helium in the core is used up, we have a repeat of the cycle in which the star implodes even further and drives temperatures up until the star reaches the billion-degree temperature necessary to fuse carbon into oxygen. And after this cycle ends, the star fuses the oxygen in the core into silicon, and so on through the heavier elements.

This process comes to an end when there is only iron in the core of the star. Iron does not release any energy when it fuses. When the star comes to a core of iron, the energies that had been pushing out from the center are now gone. There is thus nothing the star can do but implode upon itself.

In a matter of seconds, the entire core of the star becomes a tiny speck. First, all the nuclei are dismantled into their constituent protons and neutrons. Not only has the core of this once brilliant star been reduced to a speck, but the star's creativity in bringing forth these various elements is erased. And still the contraction continues. The energy of implosion becomes so great that even the free electrons and protons are crushed together to form neutrons. It is at this moment that a great reversal takes place—the supernova explosion. The force of the neutrinos, the elementary particles released during the creation of the neutrons, reverses the entire movement and blasts the star apart. The superconcentrated dot of neutrons now explodes outward with the brilliance of a hundred billion stars. And as it expands, a stupendous new round of nucleosynthesis takes place, creating the nuclei of all the elements of the universe. What had been a dense dot of matter now opens up into hot clouds of magnesium, calcium, phosphorous, carbon, and gold. This womb of intense creativity gives birth to the elements that eventually form our planet and our bodies. Much of the matter of our bodies passed through such an intense and vast explosion.

The supernova is the most spectacular display of destruction and creation in the universe. What are we to make of this, as our very existence—indeed, the very existence of life—depends upon it? Does it suggest that the universe, in order to create a single atom of carbon, requires the destruction of an entire star? Could it be that life is not possible without vast, mysterious, and ongoing transformation?

FOUR

Birth of the Solar System

Our solar system emerged out of such fiery transformation. Five billion years ago a shimmering cloud created by supernova explosions began its gravitational collapse into a thousand new star systems. Throughout this vast cloud, new centers of attraction appeared with an infant star, like a jewel shining at the heart of each center. One of these centers became our Sun with its eight planets—a solar system. This vast ocean of our solar system is like a womb that eventually brings forth life.

How did this happen?

In the beginning our infant Sun was completely surrounded

by hydrogen, carbon, silicon, and other elements disbursed by the supernova explosions. As they drifted through space these elements would brush against each other and begin to cohere into tiny balls of dust. Over millions of years these "planetesimals" continued accreting and growing until they were the size of boulders and then as large as mountains. Not all collisions resulted in larger bodies. Many were so violent that they tore both bodies apart. But over millions of years these planetesimals continued to absorb all the loose matter circling about. Our solar system, with its eight planets, its band of asteroids, and its one infant sun, slowly came into being.

It is remarkable to realize that over immense spans of time stellar dust became planets. In the earliest time of the universe this stellar dust did not even exist because the elements had not yet been formed by the stars. Yet hidden in this cosmic dust was the immense potentiality for bringing forth mountains and rivers, oyster shells and blue butterflies.

Such a process occurs over and over again in the unfolding of the universe: the self-assembling powers of the universe create new structures that allow new forms of creativity to emerge.

The long process from stellar dust to planets is filled with violence and chaos and yet gives rise to new portals of creativity. Even though this birthing took place billions of years ago, there are reminders of that original process. When at night we see a shooting star, that meteoric path of light streaking across the sky, we are witnessing one of the original pebbles of the early solar system finally coming to rest after four and a half billion years of circling the Sun alone.

ORIENTING OURSELVES TO THE PLANETS

Earlier peoples could only speculate about the formation of the planets and the Sun and the Moon. Yet in looking up at the night sky they had a sense that they were living amidst an ocean of energy swirling with stars and planets. They sought to orient themselves in this vast ocean by naming the planets and seeing living forms in the constellations of the stars. The deep drive to participate in the universe led to the creation of stories and myths —the planets were persons, the stars were kin, the Sun was a god.

In cultures around the world, this urge for participation gave rise to numerous efforts to map the movements of the heavens. To align with the planets was a means of grounding humans in the immensity of the cosmos. The rhythms of time and space were clarified—the calendar was established, seasonal rituals were determined, the cycles of agriculture were defined. Human life could thus be navigated by the planets and the stars—on land or on sea.

We too are seeking our path into the universe as we discover more about the planets—their movement and their composition. It was difficult for us humans to realize that we are living on a planet traveling around a star. One of the great moments of modern science was Kepler's discovery that planets moved not in circles but in elliptical orbits. This became a final step in the remarkable revelation, begun by Copernicus, that the planets moved not around the Earth but around the Sun. We are still absorbing the stunning realization that we are living in a vast solar system centered on a massive star.

It was only in the twentieth century that we discovered what planets are made of and how they were formed. There are two basic kinds of planets—larger planets that are gaseous and smaller planets that are rocky. In our solar system, Jupiter, Saturn, Neptune, and Uranus are the large planets. They have enough gravity to hold onto the lightest elements and thus they remain in the form of a gas. But they do not have enough gravitational strength to press the elements together into fusion processes that would enable them to become stars. As a result they remain gaseous, balanced between the realms of the rocky planets and the burning stars.

The smaller planets are Mercury, Venus, Earth, and Mars. When they first formed they were largely molten rock, but slowly, over hundreds of millions of years, these planets cooled. Eventually, Mercury and Mars solidified and became rigid all the way to their centers. But Earth—and possibly Venus—remained in a partially molten state. This special condition was the start of a new adventure in our solar system.

DYNAMICS OF EARTH

Earth's story is one of a planet finding a way to remain in the creative zone between the chaos of roiling gas and the rigidity of solid rock. When Earth was still in a partially molten state, gravity drew the heaviest metals, such as iron and nickel, thousands of miles into the core. These accumulated until eventually this dense iron core extended halfway up to the surface. Piling up on top of the core came matter such as iron-rich silicates and magnesium, components of the denser rocks. These iron-bearing magnesium silicates formed the middle area of the Earth, the mantle. Lastly, Earth's crust formed around the mantle. Only ten to one hundred miles thick, the crust is composed of light felsic rock, like granite, surrounded by large areas of ocean crust, formed largely of basaltic rocks, crystallized from upwelling magma. We can imagine Earth as an egg. Its inner core is like the yolk, its mantle is like the egg white, and its crust is like the shell.

This same process occurred on Mars, but it froze in this state. The amazing thing about Earth is that it did not freeze. A seething disequilibrium continued. The intense gravitational pressure and the heat generated by radioactive decay within the Earth produced a flow of magma as large as Earth itself. The heat gave rise to plumes of matter that floated to the surface and broke through as lava. As it cooled and solidified, the matter began its descent back into the center of the planet. This dynamic recycling of elements has, for billions of years now, operated as a process of renewal on a planetary scale.

The great convection cycle of rising and falling matter is what moves the crust over the surface of the planet. That the continents fit together like a jigsaw puzzle was first noticed by sixteenth century explorers, such as Ferdinand Magellan, who could draw upon their circumambulatory voyages to sketch maps of the entire planet. But it was not until 1915 that one scientist, Alfred Wegener, took the next step and hypothesized that the reason the continents fit together geometrically was because they were actually in motion and had once been parts of a single, connected land mass. Following his surmise, scientists over the next half century assembled the theoretical models and empirical data necessary to show that, indeed, the Earth's crust was in motion.

The understanding provided by this theory of plate tectonics must be considered one of the most significant in history, the geological equivalent to Charles Darwin's discovery of natural selection and the Einstein-Hubble discovery of an expanding universe. For just as the work of Einstein and Hubble enables us to hold the dynamics of the entire universe in mind, and Darwin's theory enables us to conceive of life as a single, complex narrative, so too the theory of plate tectonics illuminates for us the ways in which Earth developed its geological and topographic features over the past four billion years.

Earth has continuously woven itself into new combinations

out of the seething movements of Earth's plates. The plates would collide with one another and be forced back down to be melted and recycled in the mantle. Precisely because Earth lived in the zone between chaos and rigidity, its matter churned and crystallized into several thousand new minerals and a vast number of polymers. Many of these polymers existed nowhere else in the solar system and provided a precious portal for Earth's creativity. But that creativity also depended in particular ways on Earth's dynamic relationship with its offspring, the Moon.

THE PULL OF THE MOON

Just as a shooting star on a summer night can thrill us with its display of radiance, so too does the Moon enchant us. We feel the rhythm of its waxing and waning; we are fascinated by its luminous light; we are stunned by a lunar eclipse. We sense its mysterious effect on us like its gravitational pull on the changing ocean tides. A full Moon brings forth a flood of romance; a new Moon awakens promise and possibility.

Just as the rhythm of the Moon is embedded in the tides and in the months of our calendars, the mythic power of the Moon is celebrated in story and song. The Moon is the source of hundreds of myths ranging from the Man in the Moon for Europeans and Americans to the Rabbit in the Moon for the Chinese and Japanese.

Because of this power of attraction we have wondered,

"Where does the Moon come from?" This has only recently been clarified by scientists. The Moon's origin took place at the beginning of our solar system, four and a half billion years ago. As we described earlier, this was a time when the matter left over from the Sun's birth was accumulating into ever larger spheres, the planetesimals. The process giving rise to our Moon began when a large planetesimal the size of Mars collided with Earth, plowing right through the surface in the most violent encounter Earth has ever experienced. Some of the colliding planetesimal was absorbed into Earth. As Earth was largely molten, it quickly resumed its spherical shape.

But huge portions of both Earth and the colliding planetesimal were blasted out into space and formed a ring of lava around the Earth. The Moon and Earth, liquefied into magma by the collision, now separated and cooled. In a process similar to the formation of the planets, the Moon eventually stabilized. Because it was smaller, it froze into a fixed terrain of rugged highlands and smooth lowlands after one billion years.

Originally the Moon was closer to Earth. Earth was rotating more quickly and thus each day was only five hours long. Over some four billion years the Moon has been spiraling gradually outward away from Earth. As we view the Moon in the night sky we see it now as an ancient offspring of Earth, radiant with light reflected from the Sun, floating through an ocean of shimmering darkness.

THE SUN'S TRANSFORMATION OF
MATTER INTO ENERGY

If the Moon holds the mystery of night, the Sun empowers the day. Like the Moon, the Sun has an enormous effect on us—we seek its light for warmth and comfort. When we are deprived of it—especially in long winter months—we can become melancholic or anxious. The Sun literally lights us up.

For many cultures the Sun was considered a god—Ra in Egypt, Amaterasu in Japan. Great structures, such as those at Stonehenge in England and at Chaco Canyon in North America, were designed to observe the movements of the Sun. In modern times Claude Monet and other impressionist painters sought to capture the shimmering dance of light. The return of the Sun's light at the winter solstice and its diminishment at the summer solstice are still marked by festivities around the world.

But what is the source of the Sun's power, and how does it affect the planet? This massive burning star releases its energy in every direction, freely bestowing its light on our world. At ninety-three million miles away, we on Earth receive only the tiniest sliver of this energy. Yet all life on Earth depends upon this sliver.

In pondering the source of the Sun's power we can now reflect on something no earlier people could know. This knowledge became available in 1905, when Einstein discovered the equivalence of mass and energy. We now know that the Sun is converting four million tons of its mass into energy every second. In its core, the element hydrogen is being transformed into helium, releasing light in the process. The Sun is transforming its very essence into light. With each passing instant, more of its mass is becoming energy.

How riveting to discover that this transformative process at the center of our solar system brought forth life on our planet. Without the Sun, photosynthesis and green plants would not have blossomed forth, and other forms of life would not have evolved. Life is dependent on the roaring energy of the Sun—light becomes nourishment for the entire Earth community. This is the heart of transformational processes that pervade the universe. We see it in the vast explosion of the supernova; we observe it in minute chemical transformations. It is what the Chinese call the fiery furnace of the cosmos.[1]

ATMOSPHERE AND OCEAN

The fiery furnace of the universe displayed itself also in the early formation of Earth. Like soup in a huge cauldron, Earth cooked and cooled over millions of years. Volcanic processes released molten lava as well as huge clouds of dust particles and water vapor into the atmosphere. Earth was struck by large and small planetesimals, which brought more water and other compounds into the roiling mix.

The temperature of this early atmosphere was so hot that the rains turned to steam and dispersed as water vapor far above the ground. When the water eventually could reach the surface of the Earth, it formed lakes and ponds, but these quickly boiled back into steam. Earth was a fiery cauldron, in which its elements moved freely and quickly between solid, liquid, and gaseous states. This was a time of wild, frenzied activity—volcanoes rising up from the bottom of the oceans and boiling with lava, huge waves churned up by the tidal force of the nearby Moon. The oceans were a deep brown color; the sky was a pinkish-orange from an atmosphere rich in hydrogen sulfide.

As Earth continued to cool, the steam that rained down for millions of years eventually covered the surface of Earth with an ocean of water. But then a giant asteroid would once again smash through the ocean and the fragile crust, heating up the planet so that the rock melted into liquid and the ocean water boiled back into steam. Water was constantly transforming and being transformed.

Dust from asteroid impacts as well as from volcanic explosions blocked out the Sun. Night covered the Earth for millennia, until massive torrents of rain brought the dust back to Earth and a new ocean formed.

After millions of years, more stable conditions emerged for rock, water, and air. Earth became encircled by great tidal oceans and was held by a thin layer of atmosphere. Within such a double embrace of oceans and atmosphere, Earth brought forth a new marvel—the living cell.

FIVE

Life's Emergence

What did cells give rise to? We look about us now and see trees growing from the forest floor, hawks sailing through the sky, and whales breaking the ocean surface. We see the dandelion puffs floating through the summer stillness. We watch the blackberry vines grow heavy with fruit while elk fracture the night's peace with their fierce fight for mates. Such patterns of multicellular life arose over hundreds of millions of years, while antecedent forms of unicellular life arose over billions of years. During all that time, nothing similar took place on the other planets in our solar system.

On Earth, the first simple cells appeared some four billion

years ago. After another two billion years the more complex cells, those with nuclei, began to appear. This in turn gave rise to algae in the oceans, lizards in the streams, wolves in the mountains, and primates on the African savannah. The drama of life was playing out on this one remarkable planet.

How did it happen that Earth exploded with this mystery we call life? No one knows. Various theories have been proposed. Currently emerging in the sciences is a new perspective that points to the deep sea vents along the spreading centers of the ocean floor. There, the earliest forms of life were the thermophyllic bacteria that used heavy metals as their food source to sustain life in these extreme environments. With these discoveries, we are realizing that Earth's life is a manifestation of the deep patterning of the universe, referred to as self-organizing dynamics. These insights have their origin largely with Ilya Prigogine, who in 1977 was awarded the Nobel Prize in chemistry. The work of Prigogine and his collaborators provides a perspective by which the universe can be understood as suffused with a vast array of these dynamics. And when the conditions are right, any one of these dynamics can be drawn forth.

A simple example of self-organizing dynamics is the whirlpool. This particular pattern can appear anywhere so long as a body of moving liquid is at hand and some barrier is in place to disrupt and pattern the flow. The patterned flow, once established, can persist through time, although the molecules of matter are constantly flowing in and out of the whirlpool. Such structures endure so long as the flow of liquid persists.

One of the most compelling ideas concerning these structures is their nested nature. Perhaps the largest dynamic structure in the universe is that of a galaxy. Once a galactic system has been evoked, we can find self-organizing stars within the galaxy. And once we have stars, we can find self-assembling planets such as Earth, with its own organizing substructures such as hurricanes or whirlpools. Then and only then is there the possibility for a new self-organizing system, a living cell, to come forth.

This system is like a person exhaling on a cold winter morning. The first large swirls of breath generate their own smaller swirls. Then, with each passing moment, still more delicate swirls emerge within these. Such is the nature of life. The universe began with a great outpouring of cosmic breath that has complexified over billions of years until it could burst forth as evergreen forests and alpine meadows, with their cicadas and fireflies.

CELL MEMBRANES AND AWARENESS

What leads to such complex swirls of energy? Is it just random?

For centuries scientists have attempted to explain the universe by means of physical laws expressed in mathematical equations. The universe was thought to consist of mechanisms within mechanisms. Consciousness was seen as isolated in the human.

But from the new perspective of complexity science, these self-organizing dynamics can be considered the very foundations of sentience itself, for they can be understood as the processes that give rise to macro-scale physical structures such as galaxies as well as to the subtle processes of consciousness. These dynamics are something like the innate ordering processes of the universe itself.

In a simple but elegant form, awareness appears in unicellular organisms. The capacity for discernment resides in a thin outer layer of each cell, called its membrane. The membrane, through its receptor and channel proteins, selects what is of interest and what is not, what will enter and what will not. Each cell encounters a wide spectrum of atoms and molecules and other organisms floating alongside it. Each time the cell makes contact, primitive discernment emerges.

In the vast majority of these interactions, the membrane remains tightly sealed in order to block a novel molecule from its inner life. However, in encounters with molecules of particular configurations, the cell responds very differently. The molecules of the cell's membrane latch onto this new molecule. The cell then alters the structure of its own membrane so that this molecule can be drawn in. Because of this discernment, the new molecule becomes part of the cell's internal milieu. In this way the cell finds and captures its "food"—the energetic molecules it can digest.

Discernment is crucial. Mistaken decisions can lead to death because the inner coherence may be broken by the strange new guest-molecule. Thus, at the edge of its body, each cell makes an elemental choice. Is this a risk worth taking? Is this food nourishing? Will this increase the chances of remaining alive?

PHOTOSYNTHESIS

To commune may be one of the deepest tendencies in the universe.

Our planet is a riot of such communion, beginning with its gravitational relationship with the Sun. Earth has been revolving around the Sun for over four and a half billion years, and this is now a profoundly stable relationship. But the universe is not satisfied with stability alone. Over four and a half billions years, Earth has moved toward ever-greater complexity and interconnectedness.

After the emergence of life itself, one of the most stunning manifestations of this deepening communion is that of photosynthesis. The key construction, requiring perhaps tens of millions of years, was a molecular assembly capable of an elegant resonance with sunlight. Like tuning forks shaped to vibrate in the presence of certain sorts of music, these special molecules, called chlorophyll, glow with energy when the light from our Sun falls upon them. The photons, when captured, lift electrons to a higher energy state, which immediately sets off a cascade of chemical events leading to the creation of the powerhouse molecules within every cell. Life thus found a process of feeding upon the

Sun in a direct way, drawing in sunlight and using its energy to synthesize its component parts.

How can we picture the process by which photosynthesis was brought forth? To use the analogy of an engineering project, imagine a group of cells constructing photosynthetic molecules the way a group of humans would construct a bridge or a building. Thinking in these engineering terms is natural to us, because we ourselves use our hands to manipulate matter and our brains to work out a plan of action. Thus we easily and even unconsciously project such activity onto nature. Moreover, because we modern humans live in a world permeated with human-made machines, the image of the machine inevitably would dominate our imaginations. Even a genius like Isaac Newton imagined a God who organized the Heavens as a vast machine with a predetermined blueprint. But to use the machine as an image for nature's creativity diminishes the insights into life that Charles Darwin bequeathed us.

One clue as to how we might construct a better image for nature's creativity comes from the primary actors in the photosynthesis story, the primitive organisms in the oceans at least three billion years ago. We need to remember that the work of bringing forth these photosynthetic molecules was a feat of unparalleled creativity. To give a sense of how elegant these molecules are we need only bring to mind the effort scientists have expended in the study of them. Though many astounding details have been discovered, scientists have not yet captured all the marvels contained within them. It is all the more astounding when we remember that this feat of inventing photosynthesis was carried out by primitive organisms that did not even have brains, let alone eyes, hands, or libraries.

It is extremely difficult for us humans to take in the fact that nature's way of producing forms is different from that of an engineer constructing something with a blueprint in hand. Natural forms are not assembled in such a way. What we can say with some certainty is that nature is creative. And the forms of creativity that pervade nature are neither haphazard nor determined, but are, rather, profoundly exploratory, capable of bringing forth such a display of magnificence that it endlessly evokes our wonder.

So, how can we hold all of this together when we think of nature's creativity? What metaphor—what poetic image—will keep us from regarding nature as something simply engineered?

As a poetic analogy to nature's groping creativity, consider an infant's development. With a creativity that is largely unconscious, the infant over time makes a series of adjustments to her situation, adjustments that will determine much of her life. For instance, if her mother speaks Chinese, she will eventually speak Chinese. If her mother speaks Spanish, the infant will learn that language instead, which means that she will be (unconsciously) sculpting her brain and shaping her facial muscles in particular ways that depend upon the language she is learning. Similarly, through a process of trial and error, she will intuitively shape some of her basic behavioral patterns according to her mother's personality type, because on a primal level her continued existence depends upon entering into a harmonious relationship with her parent.

In the best of situations, the mother bathes her infant with love, day after day. The infant is filled with an unthinking desire to be and to live. Even before consciousness has stabilized within her, she forms herself in a way that enables her survival. In relationship with her parent, she changes her brain, her body, and her consciousness, and she moves more deeply into a stream of nourishment helping her to unfold.

So it was when Earth emerged. It did not have an engineer's blueprint. Earth simply came forth in relationship with the Sun and its intense flow of energy and began to change. We will probably never know how many molecules it tossed forth before there emerged those amazing chlorophyll molecules now in the leaves of every tree on Earth. Perhaps billions of different molecules were constructed as many different types needed to be tried. Earth would continue to change itself until this blazing flow of energy penetrated its matter. Before words, before brains, and before consciousness, there was the deep desire to exist, and the eventual discovery that it is only through relationship that we survive.

LIVING EARTH

One of the great mysteries in the evolution of the universe is the emergence of a whole out of its many parts. A star is very different from an atom, yet a star arises out of a vast cloud of atoms. Similarly, a living cell is very different from a molecule, yet a cell emerges out of a gathering of molecules.

One of the most spectacular examples of emergence is that of the integrated Earth system, our fabulously interconnected planet which arose into being out of its trillions of living and nonliving components. Some scientists go so far as to suggest that Earth itself is alive, that it is actively coordinating the temperature of its atmosphere or the salinity of its oceans. At the very least we can say that we have not yet discovered any other place with the complex coherence and interconnectivity of our planet.

The first step for this integrated whole to arise was the spread of life throughout the planet. Cells with their powers of adaptation soon filled the depths of the oceans. They lifted off and became part of the atmosphere. They emerged from the water and covered the continents. They thrived in icy snow as well as in temperatures hotter than boiling water.

At some point in its expansion, life became not just a planetary stowaway but a collaborative partner with the atmosphere, oceans, and continents in the shaping of things. The chemical composition of the oceans and the atmosphere, for instance, has been profoundly shaped by life's activity.

After life had seeped into the functioning of the planet's systems, a great emergence took place. A living planet—a complex, self-organizing system—arose with the capacity to maintain the delicate conditions of life.

For instance, the surface temperature of Earth cannot vary too far in either direction, or life will disappear. Our planet's average temperature was thought to be a fortunate consequence of being just the right distance from the Sun—ninety-three million miles. Thanks to the twentieth century discovery of nuclear fusion and the structure of stars, however, we now know that over the past four billion years our Sun has increased its temperature by nearly 25 percent.

The astonishing implication of this is that Earth has adapted itself so as to remain in the narrow band that enables life to flourish. By drawing carbon dioxide out of the atmosphere via photosynthesis, Earth altered the composition of its atmosphere to keep itself cool as the Sun grew hotter. This adaptive dance between life and nonlife changes our thinking about our planet. Earth is not just a big ball upon which living beings exist. Earth is a creative community of beings that reorganizes itself age after age so that it can perpetuate and even deepen its vibrant existence. This dynamic of reorganization is possible because of life's most essential capacity—its power to adapt.

SIX

Living and Dying

The ongoing deepening of life's complexity happens because life is able to adapt to a vast variety of conditions and to remember these adaptations, sometimes for billions of years. Nearly everything of fundamental importance in life depends upon the power of adaptation and of memory. Everywhere we look we find evidence of this process. Consider our foods. Grains, for example, are composed of many different sorts of complex organic molecules. When we eat them, they need to be carefully broken down and then woven together in a new way if they are to become part of our bodies. This complicated physiological process was worked out by trial and error hundreds of millions of years ago by cellular ancestors who are now long gone. But their accomplishments were not lost. They were remembered. As we eat, the grain is transformed into our skin, our muscles, and our organs only because life remembers its central achievements.

One way to think about the nature of life's memory is by using the insight of Pythagoras. Pythagoras's central conviction was that the essence of the universe is not water or air or fire or anything concrete like that. The essence of the universe is number; the heart of the universe is revealed in pattern.

This can seem like such an odd idea. Life is so sensually luscious, how could its fundamental nature be something abstract like a number or a pattern? The traditional story is that Pythagoras was deeply impressed by his discovery of the relationship between harmonious sound and number. Sensuously beautiful music depends on a particular ratio of vibrations. Alter this ratio in any way and the quality of the music degrades.

GENETIC MUTATION AND NATURAL SELECTION

It is precisely this deep connection between life and pattern that enables life itself to remember its crucial accomplishments. That is what the genes of the DNA of our bodies do. In their precise sequence of nucleotides, genes hold the patterns of life. The grains we eat are transformed by many proteins, where an essential player in the overall process is known as cytochrome c. Cytochrome c was assembled for the first time billions of years ago and is inside our bodies now because the information for its construction is held in the genetic patterning of our DNA. Life did not carefully hand down the actual molecule through the generations. Instead, life handed down its essence in the form of a pattern of nucleotides. Using this pattern, our bodies create these proteins anew, which then enable us to transform the grains of the fields into our flesh and blood.

The grandeur of this event is easily missed because our consciousness is not instinctually aware of the processes involved or of the effort and energy that have been poured into these processes. We use a simple phrase like, "Cytochrome c was invented . . ." and unless we put time into understanding what this means, we easily fall back into our default ways of thinking. We imagine a biological factory somewhere in the past where this one particular molecule was first designed, after which the team of engineers moved on to the next molecular challenge. What actually takes place is far more interesting.

Life's capacity to adapt depends upon the occurrence of random changes in the DNA molecule. Different patterns of nucleotides appear by chance, which lead to different proteins within the cell. Possibly millions of such proteins were generated in this way before one molecule, later named cytochrome c, enabled its possessor to survive, which led to the genetic patterns for cytochrome spreading throughout the population. This two-step process—where a vast number of trials are conducted and where the successful models can be remembered genetically—is what enables us to calmly munch on a slice of bread and transform it into the tissue of our hearts.

Though life's creativity is a groping and sometimes chaotic process, it is also a learning process. The connotation in modern English of the verb "to learn" is that of an individual acquiring a new skill. But with the discovery of biological evolution, we have a new insight into the way the ancient process of evolution can be understood as a higher-level form of "learning." We can begin with a simple question, "Who has learned to transform food into flesh?" We humans certainly had nothing to do with the construction of the physiological processes involved. Nor can we think of the early bacteria that first generated the cytochrome c protein as having any idea of what their invention might one day be used for. No, it was not any individual who learned this. It was rather life's whole process of adaptation and memory that was responsible for this new ability. It is life as a whole that learned to digest its various foods.

When we today remember that the energy of our lives comes from the original flaring forth of the universe, and that the atoms of our bodies come from the explosion of ancient stars, and that the patterns of our lives come from many ancestors over billions of years, we begin to appreciate the intricate manner in which life remembers the past and brings it into fresh form today. Life adapts. Life remembers. Life learns.

SENSING AND SEEING

We have such difficulty absorbing the magnitude of the vast amount of adaptive information that life employs because our human life span amounts to a tiny fraction of cosmic time, approximately a millionth of 1 percent. Our great challenge, then, in comprehending the universe is to overcome our natural bias that the world has always been the way it appears to us. And that it will always be more or less the same as it is now.

We can begin to appreciate something of the changing nature of the universe when we realize that even our means for sensing the processes of the universe are part of these processes as well. The way we see, the way we hear, the way we feel—each of these senses has been drawn forth and deepened for hundreds of millions of years. We see only because the Earth has long been inventing the sense of sight. And this process is not yet done.

Even certain kinds of bacteria, the first forms of life, developed sensitivity to the presence of light. But the earliest form of an eye for which there is any fossil record is that of the trilobites some five hundred million years ago. The trilobites constructed an eye using the mineral calcite. Their visual organ was a bundle of calcite rods, each rod capable of passing light down its axis without refraction. Thus the trilobite was able to see in the direction of each of these rods, a primal form of seeing that proved so successful we find it even now in the compound eyes of flies and lobsters.

An entirely different form of seeing was developed independently by worms and by fish. This eye was formed not with a hard mineral but with water, and is a form of seeing we know so well because we humans share a common ancestry with fish and thus have the same kind of eyes. What is astonishing to realize is that neither form of seeing is superior to the other. Each has its unique qualities—the compound eye is better suited for drawing light in the violet and ultraviolet range, whereas the water-based eye is better at drawing in light in the red range. This variety in the sense of seeing is found as well in all of the other senses that have arisen during the course of biological evolution.

With the emergence of the various senses, life is groping forward in an effort to see and taste and touch the world. No matter how advanced the sensory organ, the universe is never done—for there is always more to see, always more to hear. Evolutionary biologist Ernst Mayr estimates that complex eyes have been constructed, independently, at least forty times since life began.[1] Nothing will stop life's quest to absorb ever more of the universe's infinite depth.

We do not enter a finished universe. We do not enter a completed form of seeing. Scientists have articulated the details of evolution and because of this we can, using our imaginations, now begin to "see" back in time. We might call this a four-dimensional way of seeing. Once we are filled with such knowledge, our eyes can look at a honey bee not only as a small buzzing creature, but as a particular wave of life that includes the trilobite's great dramas half a billion years ago.

The early eyeless worms lived in the sandy bottom of the shallow oceans for many millennia. But then they invented a way of entering a much larger world, a world that included visual information from events hundreds of yards away from their small worlds.

Such is the nature of our moment now. Humans have lived in various civilizations such as Imperial Rome or Han China, and in each case the citizens regarded their civilization as "the whole world." But we have also discovered a new kind of "eye." With conscious self-awareness, we have developed a new kind of sight—insight into deep evolutionary time. Our vision now extends back through billions of years of evolution. With this new and powerful way of seeing, we find ourselves blinking in a thrilling and yet unsettling light. Rooted in the center of immensities, we open our eyes and see each thing ablaze with billions of years of creativity.

MULTICELLULARITY AND MIND

One of the most spectacular creations of life is the animal mind. For example, what gives rise to a lizard's mind? We can watch a lizard scamper and dart about with such intention, so different from the rock it perches on, even though they are both composed of the same chemical elements. How is it that the lizard has this hyperactive personality, or any personality at all?

We need to focus on the fundamental patterns of the lizard, in particular upon the DNA patterns organizing the billions of cells of its body. If we allow our imagination to carry us back through the mating pairs of its ancestors, eventually we come to the lizard's amphibian ancestors and then to its fish ancestors. If with our imaginations we continue following the path of sexual reproduction we will come to the most primitive animal ancestors, and when we go even further back we will arrive at the time when all of life is unicellular. We reach a time when there are no animal bodies anywhere on Earth, only various groups of squirming unicellular organisms.

This is one of the great moments of creative emergence in the life story, when something entirely new, an animal mind, appears in the evolutionary process. The mind of every fish emerged out of a long journey that began hundreds of millions of years ago in a time when there were only free-living, independent cells. Biologists speculate about this complex process now buried in the depths of time. Perhaps a single cell replicated, and instead of separating into two separate cells, the mother and daughter cells stayed together. Perhaps these two replicated in a similar manner, so that soon there were four such cells still remaining together, then eight, then sixteen. If this little ball of cells could find a way to remain together and survive, it would be entering a new pathway of exploration, one involving a group or community of beings.

As we saw in discussing the birth of life itself, we see here again the work of self-organizing dynamics. With the appearance of the first cell, we contemplated an early group of molecules, held together in a skein of chemical interactions, which evoked the patterns of activity we call life. In a similar way, this little ball of cells, stitched together in the challenges of survival, brought forth the patterns of activity we call "animal." The power of complexity that organizes the flow of chemicals between the cells is the basis for something new in the universe—an animal's mind.

Such moments of radical emergence give us an insight into what it means to be alive in the universe. Let us return to the moment a billion years ago when that little ball of cells was one of the communities making the transition into the first animals. We can reflect on the awareness, however minimal, in each of the cells. Certainly they did not have the slightest idea that their actions would become central to the process of bringing forth elephants and eagles. No matter what level of awareness each cell might have had, it is beyond imagining that they knew their work would become central to the construction of the brains and personalities of dolphins and gorillas.

It is the same with us. With all our theories and mathematics and computers, life's creativity is beyond our control. Moment by moment, we can have, at most, only a glimmer of the full significance of where our actions are leading us. We are enveloped in something like a dream. And today we are beginning to imagine that we might have a particular role to play in this dream. With each passing decade, the life process is increasingly affected by the influence of human consciousness. Perhaps human consciousness has a much larger significance within evolution than earlier philosophers could imagine. Could it be that our deeper destiny is to bring forth a new coherence within the planet as a whole, as the human community learns to align itself with the underlying dynamics of Earth's life?

VIOLENCE, DESTRUCTION, AND DEATH

In our search for this planetary coherence one of the most difficult elements of the universe to navigate is its violence. Humans have, from earliest times, been challenged with the task of orienting themselves to destruction and violence. Some religious traditions have seen the world as necessarily violent, as a battleground for the forces of good and evil. Others have convinced themselves that we can change the world, that it is only a matter of time before we improve things so that eventually there will be little or no destruction.

If the discoveries of the universe and Earth stories have introduced us to unsuspected grandeur and magnificence, they have also brought into our awareness dimensions of destruction terrifying to contemplate. Stars have exploded with such violence they blasted any nearby living planets into swirling clouds of dust. Or, even on the much reduced scale of our solar system, there is the real possibility of a collision between an asteroid and our planet that would send billions of sentient beings into meaningless agony and death. Contemplating both the possible and actual suffering in this universe, one can easily understand the continuing appeal of nihilistic or narcissistic philosophies. Surrounded by the absurdity of needless suffering, why not devote one's energies to one's private pleasure and success?

The discovery of Earth's evolutionary processes does not answer the deep questions of violence, but it does suggest other possible orientations toward violence and destruction. For example, rather than demanding a rational explanation of destruction, or insisting on eliminating all violence, we might aim instead at orienting ourselves creatively in the midst of destructive processes. Such an orientation is powerfully present in nature, especially in the predator-prey relationships found throughout the natural world.

When we observe a lizard and the amazing way it blends into its habitat we can find ourselves admiring its struggle for survival. So much so that when a hawk comes screaming out of the sky and kills the lizard we might wish the hawk would find other food. But we can as easily take the perspective of the hawk. If we become fascinated with the beauty of the hawk's form and the penetrating intelligence of its eyes, we can find ourselves hoping that the hawk captures its lizards and whatever else it takes to keep its magnificence alive.

As Charles Darwin realized in the nineteenth century, what is astonishing is that the elegance of the hawk depends upon the lizard. And the elegance of the lizard depends upon the hawk. Their qualitatively different attributes arise in tandem.

It is the drama of survival that presses the hawk to become ever more astounding in its capacities for flight. Each generation of juvenile hawks enters the struggle with slightly different capacities. Because the lizard is difficult to capture, the more impressive hawks will have better success at staying alive and will thus pass on their skills to greater numbers of offspring in the next generation. But any such advance in the hawk's hunting skills will set up the conditions leading to a corresponding advance in the lizard population's capacity for escape, a complex process known as coevolution.

Though violence stuns our rational minds and defies our attempts to explain its existence, we do learn from life that destruction is inextricably entwined with the process of bringing forth ever more complex forms of life. Though we are not able to eliminate violence from the Earth or universe, there is the possibility that we will move forward in ways that diminish the destruction. And we may even learn to orient ourselves with its presence in a manner that is creative or life-enhancing.

But what of death? Death presents us with one of the most fundamental challenges to the human spirit. Death of a loved one—especially a child—confounds our efforts at understanding.

We pay respects in memorials for the dead and in visits to ancestral graves. These occasions draw us together in the presence of suffering and loss, assuaging personal grief and channeling it into a shared communal experience. But they do more than this.

Such rituals often situate individual death in relation to the great cycles of nature. They place us within the vast community of the living and the dead and in so doing we enter into the processes that nurture future life.

But we still ponder the question, Where does our life energy go? Does deep time, geological and cosmological, offer any insights here? If we have come out of this immense journey is it not possible that our transition is a return?

Can it be that our small self dies into the large self of the universe? Are our passions and dreams, as well as our anguish and loss, woven into the fabric of the universe itself?

SEVEN

The Passion of Animals

P assion—our urge to merge. What is more intimate to our souls? Our passions determine so much of our lives. They are the wild, explosive energies of love and creativity. They inexorably shape us into the sorts of people we become. Desire lives at the heart of life's evolution.

The ancient Greeks symbolized desire as a gift from the Gods, even as a visitation from one of the Gods—Aphrodite or Dionysus. Their imagination expressed their awareness of the power and significance of this emotion. In many of the Greek myths, the characters are portrayed as utterly at the mercy of desire.

From our own scientific perspective, it is remarkable that we too have come to an understanding of the centrality of desire and longing in the evolution of life. One of the most amazing discoveries is that life has arrived at ever deeper passions and ever more effective ways of satisfying these passions. We begin to sense this deep pattern in the universe when we reflect upon the arc of vertebrate evolution from fish to reptiles to mammals.

The male stickleback fish, with his red belly and blue eyes, expresses his desire by his courtship dance. If he is successful, his mate will deposit her eggs in his nest and he can quickly move in and fertilize them.

More than one hundred million years later, when lizards have evolved from their fish and amphibian ancestors, the passion to merge has deepened. We see this most profoundly in the invention of sex organs in the reptiles, the vagina and penis. Thus the reptiles are able to consummate their desire by intermingling their bodies in a way that the stickleback fish cannot. After years of living in separate destinies, two reptiles merge into a single communion experience that changes the course of evolution.

With mammals, the passions reach yet a new crescendo. Not only are they able to commingle as one body, they can become so profoundly bonded that they remain in relationship their entire lifetimes. It is as if in the secret heart of life there awaits an infinite delight, an ecstasy dimly perceived by the earliest fish and now flowering forth among the mammals. As D. H. Lawrence wrote of two whales mating in his poem "Whales Weep Not!":

And they rock, and they rock, through the sensual ageless ages
on the depths of the seven seas,
and through the salt they reel with drunk delight
and in the tropics tremble they with love
and roll with massive, strong desire, like gods.

ANIMAL MATING

All animals live in the great drama of their passions. Consider the male bowerbird. On awakening each morning in spring he applies every scrap of surplus energy to building his nest. He gets twigs and fastens them together. He scours the area for the perfect building materials. He arranges and then rearranges things just like a carpenter obsessing over a new building. He will even make a dark paste and use this to paint some of the walls of his house.

How does he know how to proceed? Some of this knowledge he learns from his own trial and error, but it comes largely from the patterns stored in his DNA by his most clever ancestors. They were striving for the same thing. They were attempting to become excellent in the same way, the way of extravagant architecture that their species had settled upon.

The spiders, on the other hand, chose rhythm—males learn to play the strings of the web as if they formed a primitive lyre. Dance is another channel for communication—the stickleback fish learned to dance underwater while the cranes learned to dance on land. Even anatomy is used to impress a mate, such as the peacock, whose tail feathers rival the beauty of the rainbow.

What is true of each of these males? Why does he throw himself into such activity, all of it costly and some of it life-threatening?

He is seeking to convey his deepest truth—that he finds her valuable. Life has shaped his mind in a particular fashion. He cannot see all the value in the universe, but he can see hers, and it might as well be infinite, for nothing matters in comparison. His great passion is to organize his life around the work of wooing her, of impressing her, of changing himself in whatever way he can so that she will look at him with admiration and will utter in her own wordless way the longed-for magic contained in that one word, "Yes."

FEMALE CHOICE

Then again, she might not say yes. After all his hard work, after displaying all his skill and talent, her response might simply be, "No, thank you."

We live in a universe where choices are made. Not everything goes. Some things are better than others and are preferred, which inevitably causes strife. For everyone wants to be the chosen one; everyone wants to be the one of great value. No one wants to be refused or tossed aside. But life's selection dynamics can be harsh.

No matter how difficult it might be to choose one and reject the others, life insists upon it.

In the animal world, sexual discernment is usually carried out by the females in a process Charles Darwin called sexual selection. Thus the male bowerbirds construct their fancy nests and the female birds fly about and choose among them. Similarly, it is the male sticklebacks who change their coloration and perform their dance, but it is the female stickleback who decides yes or no.

We see in this interaction of the genders the same sort of disequilibrium that has appeared elsewhere in the universe, as in the stars with their explosive and implosive forces. Here in the gender dynamic the tension comes from two polarities: on one side is a male proclivity for widespread mating, and on the other is a female insistence upon discerning a high-quality male. What is intriguing is that these poles are not there accidentally or mistakenly; they are there because self-organizing powers require this tension.

It is easy to see how life rewards the male polarity. If a male has the genes that urge widespread mating, he will mate, on average, with a much greater number of females than will a male who is very particular. In the next generation there will thus be a larger number of males carrying the genes for widespread mating.

But the same strategy does not work for those female animals, such as mammals, who give birth after a period of pregnancy. Such a female generally puts a much larger investment of energy into both the pregnancy and the rearing of her offspring. So her chances for influencing the future depend hugely on the quality of her offspring. She has only one litter per season, whereas an indiscriminate male might have dozens. Thus, life pressures her to choose the mate with the highest quality or vitality or health.

Of course, when we use words like "choose" or "decide" we need to remember that such actions are not exactly the same for birds as they are for humans. When a peahen beholds the trembling display of a large peacock, she is probably not making a sophisticated mental list of attributes before arriving at a decision concerning this suitor. Rather, her ability to evaluate is deeply woven into the very functioning of her complex awareness. She simply knows, directly and intuitively, because for millions of years life has been shaping her ancestors' sensibilities to notice and prefer those qualities in the male that have the highest probability of ensuring a secure future for her and her offspring. The capacity for identifying value has been worked into the foundations of her mind.

The males in their deep passion pour forth an extravagant display of promise and beauty; the females in their perspicacity sift through them all and make life-and-death decisions. And as each of us is a mingling together of genes from a female and a male, these ancient capacities of desire and discernment are layered into our flesh and bones.

PARENTAL CARE

We find some of the most universal themes of feminine wisdom in such celebrated icons as Mary in western Europe or Kuan Yin in China or Oshun in Africa. Do such images from all around the planet relate only to the human? Are they disconnected from the larger processes of evolution? Or is there a way in which these images, with their celebration of compassion, relate directly to the dynamics of the universe?

"Parental care" is the phrase ethnologists use for those activities on the part of a parent that are concerned with the well-being of its offspring. Such behavioral patterns are often contained within the genetic programs of the mother, but they can also come from cultural patterns within a particular population. We find such caring behavior throughout the mammalian worlds where parents, especially mothers, devote time and energy to the nurturance and education of their young. Mammalian mothers—whether lions, chimpanzees, or deer—dedicate days, weeks, months, and even years of their lives helping to bring forth what will become the next generation of their species.

Such caring behavior is not a unique invention of the mammals, for we find similar traits among the reptiles. Even though the amount of energy devoted to their offspring is usually not as extensive as it is for the mammals, nevertheless many reptilian mothers express a concern for their offspring by remaining with the eggs after they have laid them, curling around them to keep them warm and thereby increasing the probability that they will hatch. And when the babies do struggle out of their eggs some mothers will stay with them to scare off predators.

How far back in the evolution of the animal world does this caring behavior go? No one knows. Paleontologists have found fossil evidence suggesting that such care was exhibited by the dinosaurs a hundred million years ago.[1] And there are reasons to wonder whether the early fish possessed similar traits. Certainly we can observe ways that some contemporary fish express parental care. Mothers will remain in close proximity to their fry and aggressively attack any fish that might approach. This particular trait is not found universally among fish, for many mother fish will unhesitatingly feed upon their own offspring, but we can speculate that the trait found in some present-day fish had its origin even hundreds of millions of years ago.

All of these examples come from the animal world, where we can observe behavior that is easily identifiable as caring. But such an approach might be too limited. Consider the amount of time and energy a Douglas fir tree will invest in the process of making its cones. Each year the fir trees of a temperate forest will create literally millions of pine cones, all of which are dispensable for an individual tree's functioning or survival. The pine cones are constructed to bring forth offspring. It goes without saying that the parent trees have no consciousness of the meaning of their work in creating these cones. But whether or not there is any awareness of what they are doing, the fir trees' actions can be understood as a kind of preconscious concern for the next generation.

We can extend this reasoning to the most primitive form of life we know—prokaryotic organisms. The most complex work these prokaryotes do is to create new daughter cells. From the perspective of a single prokaryote, a vast amount of matter and energy is expended to create something that has nothing to do with its own survival and everything to do with the survival of offspring. Thus, in the sense we are developing here, a concern for the well-being of offspring can be understood as something that is woven into the very fabric of life itself. From this larger, cosmological perspective, the images of Mary, Kuan Yin, Oshun, and other goddesses can be understood as being in alignment with the same ancient dynamic of life's concern for its offspring. Over billions of years, life has developed myriad ways of expressing this care or concern, and of remembering these, primarily via the DNA. But with the human species, a new way of remembering and amplifying compassion has emerged—a means involving something never seen before in the history of life. In order to understand this new dynamic, we must take a step back and reflect on the emergence of the human as a whole.

EIGHT

The Origin of the Human

W hat gave birth to the human?

Our current best evidence suggests that something of profound importance took place five to seven million years ago in Africa. Something happened that ignited the human lineage of the primate world. A new line of energetic apes emerged that would, over the next several million years, bring forth massive brains and learn to dwell in a world saturated with dreams. Nothing like them had ever existed before. So what was it that gave rise to them?

Considerable fossil and genetic evidence has been obtained on human origins, and even though much remains to be learned, we can outline our amazing origin story in some detail. Our human odyssey began some six or seven million years ago with a population of perhaps a hundred thousand chimplike apes living near the very center of Africa. Earth's climate was changing in a way that was drying out central regions of the African continent. The forests and their abundant foods were disappearing. In response to this crisis, the ancestral population adopted two very different strategies. One portion clung to the shrinking forests in order to maintain their way of life in the midst of a difficult environmental transition. But another group abandoned its past and ventured out into the open spaces of the savannahs. It was ill-prepared for this move. Its highly developed skills for swinging through the trees amounted to nothing out there under the blazing Sun, where it was hunted by hungry predators of several different lineages, including the wild dogs and the great cats.

Assisting this new species in its struggle to survive was the emergence of new traits, principally the ability to move about on two feet. Bipedalism was certainly in place by four million years ago, for we have the evidence in Tanzania of fossilized footprints of two such early humans, perhaps a parent and a child. The second major development was an increase in brain size. The first hominids had brains the size of an orange, four hundred cubic centimeters. By the time the brain had reached the volume of a grapefruit, some eight hundred cubic centimeters, humans had mastered the ability to make stone tools. And then, only six million years after these human primates had made their daring move into the savannahs, their brains had reached the size of a melon, some fourteen hundred cubic centimeters. It was at this time that an entirely new adventure was about to begin.

It was more than fifty thousand years ago in northeastern Africa that some of these completely modern humans moved out of Africa altogether. Their population in Africa was perhaps as small as five thousand, and the departing group was even smaller, perhaps as few as a hundred fifty humans altogether.[1] This small group crossed the Red Sea at its southern end to enter what is now the Arabian peninsula. Some of their descendants worked their way along the coastlines of India. Others turned north into what is now Europe, where great cave paintings remain in northern Spain and southern France. Eventually they spread east across the Eurasian continent, finally crossing the Bering Straits and moving down into the Americas.

We will most likely never know the full story of their exodus in all its drama and challenges. But what we do know is that a band of several hundred men, women, and children spread and multiplied through the centuries until this new African species was populating every continent and biome of the planet. What was it about the early humans that enabled such spectacular biological success?

HUMAN FLEXIBILITY

Though we do not have a detailed understanding of the biological changes that enabled this human species to quickly encircle the planet, biologists and anthropologists have identified some of the major factors at work. As in previous evolutionary explosions, it was a case of radically new capacities emerging out of a new set of relationships among already existing entities. If, in the early Earth, a new network of molecular relationships brought forth life, and if, some five hundred million years ago, a new order among unicellular organisms brought forth animal minds, it is equally true that over the past several million years a new complex of traits— most especially bipedalism, increased brain size, and behavioral flexibility—has drawn forth out of its primate foundation a new species now known as humanity. We have already mentioned the first two traits, bipedalism and increased brain size. It is the third trait, behavioral flexibility, which is especially intriguing.

Even though we do not yet know the genetic changes that led to behavioral flexibility, we do know that these changes were fundamental in the emergence of our species. Though it is somewhat counterintuitive to our usual way of thinking, something new began, not with an addition, but with a loss, a loss of some of our instinctual responses.[2] When these instinctual patterns of behavior were lost, a new experiment emerged in Earth's journey. Accompanying this loss, a new kind of consciousness was entering existence, one that was freer and even more exploratory. When we remember the vast amount of time that had been required for life to build up these instinctual responses, we begin to recognize the radical nature of this moment. What had been dominant was now damped down, for something different was about to come forth.

What would it have felt like, to emerge as one of these first humans? We can only guess, of course, but we do have some contemporary biological examples that bear a faint similarity to these primordial humans. The first humans would have a quality that is shared by the offspring of every mammalian species—behavioral flexibility. We find such flexibility in the young of all mammals. Mammalian juveniles resemble our earliest human ancestors in their relative freedom from the genetic constraints that will begin to govern them as they enter their adult forms.

For the young mammal, behavior is open-ended in a way that is rarer in adults. This youthful behavior is readily distinguishable from the serious adult concerns of survival or sexual reproduction. Certainly some of their playful activity can be understood as preparation and practice for their later lives. But much of it is without any direct relationship to adult behavior. In a word, what often occupies their consciousness is play. They leap and twist; they explore the world with their eyes; they taste the world with their mouths; they enter into many kinds of relationships out of sheer curiosity. With their play they are discovering the exuberance of being alive.

So, in wondering about the emergence of this new human species, we can hardly do better than to ask some simple, rhetorical questions generated by our reflections on youth. What if life could one day bring forth a species that could dwell in flexibility and resilience? What if, after a hundred million years of mammalian existence, there appeared a species that could remain spontaneous, curious, astonished, compelled to try everything? What would happen then?

SYMBOLIC CONSCIOUSNESS

Our behavioral freedom and our curiosity led to an entirely new level of consciousness. Consider a moment from early human evolution, say a hundred thousand years ago. Some gazelles and some humans are confronted by the towering flames of a forest fire. They feel almost identical sensations and emotions—the heat, the light, the roaring sound, a fear that freezes the skin, a thrill that rises up from the stomach. But there would be a difference in their response, for the gazelles will know what to do to survive, which is to flee. The gazelles will, in that sense, experience more than the humans. The gazelles will experience not only the heat and light and roar but also an irresistible urge to dash away, a strategy deeply layered into their genes and now fully activated by the forest fire.

There will be feelings of fear in the humans as well. Yet with destruction all around, with the other animals frantically fleeing, the human might instead stand transfixed in wonder. Instead of fleeing from the flames the human might even be drawn irresistibly toward them. It was this relative freedom from instinctual behavior, it has been argued, that enabled us to become profoundly captivated by so many things—by fire, sunrise, ocean waves, erotic intensities, the death of a friend, the birth of a child.

All of these events must have stunned the early humans, drawing them ever further into their experiences. They viewed life through new eyes. Instead of simply responding, they also reflected. They tasted the very essence of what it meant to be alive. With the emergence of the human, the universe created a space where depths of feelings could be concentrated and wondered over.

Humanity's greatest invention, called symbolic language, enabled humans to share this superabundant consciousness. The universe had reached a new fever pitch in the human, and this boiled over as words. It is as if the early humans were unable to contain such intensities so they constructed mental forms that would carry off some of this white-hot awareness.

Now, even years after an event, humans gathered in the twilight and made rasping sounds, and suddenly in their midst there lived again the roaring fire and the stampeding animals. These early humans huddled together as if the flames were now singeing their hair, and once again they were petrified with fear and captivated with the thrill of existence.

With the invention of the symbol, humans released their blazing imaginations into the world. Nothing would ever be the same again. With the creation of language humans entered into symbolic consciousness. Now humans could remember—could celebrate the great events of their journey. Story was born.

CULTURE AS COLLECTIVE DNA

A further step toward becoming human took place when early people learned to externalize consciousness. By creating marks on bones or in wet clay, humans invented a way to cast their consciousness into an enduring form outside of themselves—a deer antler with notches becomes a condensation of human understanding of the Moon's position in the sky.

Nothing this significant had happened in the life process since the emergence of the DNA molecule four billion years ago. DNA is life's way of storing the information of the most significant changes that have taken place in evolution. For instance, if parental care becomes hardwired into a particular fish's DNA, future generations of that species of fish have those instructions, long after the lifetime of those fish in whom the parental care mutation first arose. Thus, when some process is captured in the

DNA molecules, it can become a legacy that transcends time. But if it is not captured in DNA it is lost.

Something similar happened when humans learned to externalize their awareness in cultural forms, but now with a big difference. The great insights of humans could now be preserved in painting, poetry, and prose, so human culture became a kind of DNA outside the body.

With human culture, experience itself can be remembered and passed down. No change in the DNA is required. Any valuable understanding, even if experienced by a single human being, can become part of the enduring legacy of humanity. This is the power of language—oral or written. Such, for example, is the legacy of Chinese Neo-Confucianism we have inherited from Zhang Zai in his Western Inscription: "Heaven is my father and Earth is my mother and even such a small creature as I finds an intimate place in their midst. Therefore that which extends throughout the universe I regard as my body and that which directs the universe I consider as my nature. All people are my brothers and sisters, and all things are my companions."[3]

With the Western Inscription, as with all great literature, humans found a way to preserve the most penetrating insights so that these could become part of the enduring context into which each new human is born. A child is thus given not just the genetic treasures of its parents, but a transgenetic treasure drawing upon a culture's wisdom. Zhang Zai arrived at this vision of comprehensive compassion and universal kinship. Instead of this insight disappearing at his death, it was captured in language and thus could potentially become part of the formative influences of humans around the planet.

ENVELOPING THE EARTH

Through culture we constructed ways to share our minds and feelings with each other in story and in art. Working together, we also invented powerful strategies for survival.

We preserved our knowledge in forms that could accumulate over the centuries, for we put our stories and achievements on clay tablets and on stone. Thus every succeeding generation could draw on knowledge that gifted humans of the past had learned and bequeathed to them—epics and scriptures, poetry and plays, treatises and almanacs.

Nothing like this had existed before. This knowledge and the power for survival that flowed from it enabled humans to transcend the constraints put in place by four billion years of life. For none of the other thirty million species had the capacities of this new symbol-making animal.

Humans have at their disposal vast storehouses of learning accumulated and refined over millennia in written and oral traditions. There is little validity to the idea that humans are isolated individuals, for each of us arises out of an ocean of experience and understanding acquired by our species as a whole. As we learn to draw upon aspects of this accumulation of knowledge we begin to participate in a collective process that has developed for some two hundred thousand years.

And with this consciousness, these ever-more-powerful humans moved out of Africa and spread over the Earth's surface in a geological instant, becoming a presence on every continent. Survival in mountains? In deserts? Traveling through blizzards? Across oceans? Nothing could stop the human odyssey. Because of our symbol-making skills, we became, overnight, a planetary species.

NINE

Becoming a Planetary Presence

E very place we went, we became that place. That is the brilliant power provided by symbolic consciousness. With their cultural inventions, humans could adapt to new environments much more quickly than would be the case if they had to rely solely upon genetic changes. That's why the humans who decided to follow the reindeer rapidly became reindeer people. They walked the same pathways as the reindeer. They ate some of the same foods. At night, in their feasts and their dancing, they celebrated the thrill of being the reindeer people.

Other humans aligned themselves with the whales and became the whale people. Some identified with the birds and began wearing feathers and greeting each dawn with song—their highest fulfillment became the act of joining with the birds' celebration.

The early humans did not just journey through Earth's worlds. The spirit of each world captivated their imaginations as they re-visioned their lives in terms of that place. They absorbed every color and sound into their life and soul.

What is fascinating is that this original odyssey around the planet coincided with a wave of loss. Some humans ventured out of Africa fifty thousand years ago, reaching North America some eighteen thousand or more years ago. In another mere thousand years they were already at the southern tip of South America. But during this period vast numbers of the mammalian species on the twin continents disappeared, including a majority of all the large animals.

While it is uncertain whether this destruction was anthropo-genic or the result of climate change, it is nonetheless disturbing that our journey into a planetary presence would be accompanied by such a diminishment of life. There is both power and poi-gnancy to the human story.

The genius of the human, as displayed by this rapid movement around the planet, is the capacity of symbolic consciousness to survive in any biome. The challenge of the human, perhaps indi-cated by the widespread loss of animals, is learning to manage these seemingly infinite powers bestowed by symbolic consciousness.

HIERARCHICAL CIVILIZATIONS

Because the planet is a sphere, there came a moment when these most successful humans folded back upon themselves, meeting the one animal that could match them—their distant cousins. Thus, after nearly two hundred thousand years of life in hunting and gathering bands, humans found themselves moving into fixed settlements near the richest river deltas—the Nile in Africa, the Tigris and Euphrates in Mesopotamia, the Indus in India, the Yellow River in China, the Mississippi in North America. From these river valley settlements humans began to interact with one another in ongoing and ever deepening ways.

When the universe folds back on itself, it complexifies. We saw this earlier with the birth of a star, when the simple atoms are forced, through compression, to produce a hundred elements. The intensification of life is similar; when individual cells stayed in close connection with each other, they learned to cooperate and specialize and gave birth to the plants and animals.

The emergence of civilizations was a major event in the complexifying of Earth. The first cities—such as Jericho and Sumer in western Asia and Harappa and Mohenjo-daro in southern Asia—became cauldrons for human creativity. The possibilities for being human simply exploded. Architecture and wheeled transportation, brick making and stone cutting, literature and ceremonial art, weaponry and metallurgy, legislation and bureaucratic administration, and above all agriculture and seed preservation either were developed for the first time or were deepened significantly. All of this had a profound impact on human consciousness. Reflecting on the changes of the seasons in planting and harvesting, humans awoke to the vast transformations in Earth's processes and wove these into the yearly cycle of rites. Observing the fecundity of seeds as they grew into plants, humans became attuned to the fecundity of Earth and celebrated this in symbol and ritual.

This entire process rested upon the ability to evoke and sustain cooperation, especially for food production and sharing among diverse groups of people. There were no genetic instructions in our DNA for managing these feats. Thus, beginning some five millennia ago, human activity began to be shaped in decisive ways by language, especially by the symbolic codes of civilization.

These legal and ethical codes created coherence out of what would otherwise have been a chaos of conflicting human energies. Each tribe was forced to sacrifice some of its former autonomy, but the resultant coordination of energies irrevocably changed the face of Earth. Pyramids rose up from the African desert. Ancient rivers were diverted. Land as large as the eye could see was watered by irrigation systems. The forests scattered across the oceans in the form of sailing vessels.

With the rise of horticulture, seeds were no longer subject to the vagaries of climate, but received their watery nourishment with the precision and inevitability of logical thought. What does it mean when even the seeds begin to live not just in the Earth but in an Earth shaped by human consciousness?

It means that a great transition was under way. The geological and biological structures of Earth were becoming permeated with human presence in the form of codes, symbols, and writing.

A CLOCKWORK UNIVERSE

Maybe it was inevitable that we would become completely fascinated by our symbol systems, whether in pictographs, alphabets, or numbers. These codes became the patterns for organizing society and civilization. They constructed our food systems. They organized our militaries. They provided the design for our temples. It was only natural that we would become thoroughly absorbed in our own symbol making, eventually constructing educational systems that focused primarily not on life itself, but on language, mathematics, technology.

In addition to all this, there was something more, something that Pythagoras intuited two and a half millennia ago. There was an undeniable magic to the symbols themselves. In some mysterious way they connected us to the depths of the cosmos.

Consider Newtonian physics. Few symbolic codes have impressed the human mind more than Isaac Newton's equations for motion. From one perspective, they are just squiggly lines scrawled on paper; from another they can predict with utter precision the movements of the planets in the heavens or a stone thrown on Earth. So impressed by his equations, European philosophers such as Voltaire and Immanuel Kant concluded that Newton's work was equivalent to a revelation.

In time, many educated Europeans became convinced that the universe operated just as Newton's equations indicated it did. Matter was passive and moved according to deterministic laws that could be discovered by human reason. They imagined that these laws were established at the beginning of time by a deity who then let his world machine run forward. They called this a Clockwork universe.

THE DREAM OF PROGRESS

We modern humans became so fascinated with the power of these mathematical equations that we took these abstractions for reality. If we found the right numbers and put them into the equations, we would know how things would unfold into the future.

In the seventeenth century John Flamsteed set up his Greenwich Observatory with this in mind. Flamsteed stayed up night after night for over forty years, mapping out the positions of the Moon and stars. The hope of this endeavor, which was pursued by other leading European nations, was to obtain the precise data that would enable navigators to locate their position on the oceans.

Earth's surface became a grid filled with the numbers of latitude and longitude. That was all that was needed to control movement over the seas. With their equations and their measurements, the early scientists were discovering truths none of the classical scholars had known. In chemistry, physiology, engineering, and astronomy, these Europeans realized Pythagoras had been right all along —the world could be understood through number and pattern.

One of the defining characteristics of this new, modern form of consciousness was the decision to transform all of this symbolic knowledge into machines. A machine is a physical system that focuses energies toward satisfying some human desire. So what began as mathematical patterns within consciousness became externalized as internal combustion engines, hydraulic pumps, electric motors, and, over time, computers, controlled nuclear fission, and genetic engineering. This technological revolution became the engine of modern progress.

The human had achieved enormous powers for manipulating and commodifying matter. And for the early scientists who assumed matter was neutral, all meaning was derived from how this matter was put to use for humans.

Modern industrial humans broke with the past. They did not seek to commune with nature, or to revere it as divine gift. They sought to transform the world. For they had a dream. Using these new technological powers, they would create a better world, one with greater quantities of food, more efficient transportation, and faster communication. Using their new machines, they would eliminate poverty, hunger, and sickness. Previous peoples might have lamented earthly existence and dreamed of a heaven one entered upon death. These moderns had a different dream. They would use their power to build this heaven here on Earth.

THE END OF A GEOLOGICAL ERA

Dynamized by their technology and their dreams of material progress, modern humans transformed the planet into a bundle of resources. They produced food in quantities never seen in history, and consequently populations exploded. During Newton's lifetime a half billion humans populated the planet. At the start of the twenty-first century that number is close to seven billion and is growing. Feeding and housing this many humans have led to the depletion of the oceans, the degradation of the forests, and the loss of topsoil.

In all this, modern humans are simply extending processes begun in earlier civilizations, and in so doing the geological and biological structures of the planet are becoming ever more permeated with human presence. For billions of years such dynamics as natural selection or genetic mutation operated without any reference to human consciousness. And when humans first emerged two hundred thousand years ago, their presence was negligible compared with these larger processes of life and planet. But during the modern period a transition has taken place. Because of the power of symbolic consciousness to amplify our control, humans have begun to alter the very functioning of what was previously an entirely wild selection process in nature. We have crossed over into an Earth whose very atmosphere and biosphere are being shaped by human decisions.

We can see this very directly by looking, for instance, at the cheetah. Cheetahs attained their magnificent form through millions of years of life in the wilds of Africa, in the mysterious process of biological evolution. But today, cheetahs are no longer evolving in the wild. Rather, they live inside game parks whose location, size, and population are largely determined by humans.

A similar account could be given for the hippos, the gazelles, the zebras, the songbirds, and the sea turtles. Each of them now evolves through interactions in a world that is structured, in significant ways, by the effects of symbolic consciousness. We have thus radically altered the evolutionary dynamics of Earth.

We live on a different planet now, where not biology but symbolic consciousness is the determining factor for evolution. Cultural selection has overwhelmed natural selection. That is, the survival of species and of entire ecosystems now depends primarily on human activities. We are faced with a challenge no previous humans even contemplated: How are we to make decisions that will benefit an entire planet for the next several millennia?

We have only just awakened to this larger responsibility, and at the same time we are discovering the massive destruction taking place. With our machines and our numbers we have become a geological force. Because of us, the ice caps are melting. Because of us, coral reefs the size of mountains are dying.

We thought we were making a better and more prosperous world. But from the perspective of life, we have done the opposite. The paradox of unintended consequences is now becoming evident. The oceans, the rivers, the atmosphere, and the soil have all been severely degraded by our actions.

Nothing shows the disaster more clearly than the fact that we are causing thousands of species to go extinct each year. Indeed nothing this devastating has taken place on Earth since the extinction of the dinosaurs. Though certainly unintended, one of the consequences of the modern form of humanity is the termination of the Cenozoic era, which had its beginning sixty-five million years ago. With this current mass extinction, then, we are leaving behind the Holocene period of the past ten thousand years and entering the Anthropocene, an era shaped primarily not by natural systems but by humans.[1]

TEN

Rethinking Matter and Time

From its inception, modern science was committed to discovering knowledge and using it to make a better world. Why, then, with all of this scientific knowledge and technical skill, have we caused such extensive damage to Earth's ecosystems? For the most part, this destruction is carried out without any deep awareness that life required literally billions of years to bring forth such complexity. What is it about our modern consciousness that enables us to avoid seeing the disastrous results of our way of life?

Perhaps the destruction comes, at least in part, from an inadequate understanding of matter itself. "Deterministic materialism," as it is sometimes called, was a worldview that emerged in the late sixteenth century and seventeenth century amidst intense debates over natural philosophy. It had three tenets: that all things in the universe were composed of tiny particles of matter; that these particles were purely material, without any degree of subjectivity; and that these particles moved according to fixed, mathematical laws.

Galileo was the first scientist to adopt and develop this perspective. Following the analytical approach that he found in Archimedes, he successfully explained the movement of balls down inclined planes using mathematics alone. He left behind the ideas of Aristotle and the philosophers of the Middle Ages in Europe who tried to explain motion by reference to an inner form that each thing possessed. Galileo and the scientists who followed him ignored such qualities as form or beauty and focused their attention on quantities. Their approach led to insights concerning the natural world that no one before them had discovered.

This approach reached its culmination with Isaac Newton, who was born on Christmas Day, 1642, the same year that Galileo died. Newton extended Galileo's reasoning from rolling balls to the very planets of our solar system, and with similar success. Beginning with the assumption that Jupiter, Mars, and Venus were not gods but simply balls of dead matter, Newton articulated the exact mathematical paths that each of them would traverse, throughout the year and far into the future.

This was described with tremendous mathematical original-ity in Newton's *Principia Mathematica Philosophiae Naturalis.* Its publication in 1687 was one of the great moments in intellectual history. Even though his *Principia* became the foundation for deterministic materialism for many scientists, such mechanistic philosophy was at odds with Newton's own mystical intuitions about the nature of the universe. Mechanistic philosophy pushed to the sidelines the ancient ideas of sacred groves or holy springs. Even the Aristotelian idea that animals possessed souls was dis-carded. Spreading throughout Europe was this new idea that the universe was a vast machine run by natural laws. And the human agenda was also set anew. As Francis Bacon and others proclaimed, we modern humans with our vast intelligence had only to deter-mine the laws governing matter for us to gain control over the entire affair.

As the scientists were outlining a new understanding of mat-ter, René Descartes philosophically promoted the idea that matter is passive and inert, while the mind exists only in the human brain. For Descartes, and for many of the Enlightenment thinkers, matter came to be seen as substance devoid of subjectivity. Only humans had thought and feeling; other animals and the rest of nature operated like a machine. It is not hard to imagine how such a reductionist worldview would later provide a rationale for in-dustrial society's dismantling of Earth's ecosystems. Old-growth forests are clear-cut; robust fisheries are depleted. Ancient mountaintops are removed for coal, oceans are mined for oil. From this modern perspective, matter exists primarily for human use. The intricate living systems of nature are exploited. Such a worldview now pervades contemporary thinking.

Chemist Ilya Prigogine was one of the first to go beyond modern science's assumptions concerning the passivity of matter. Prigogine's experiments demonstrated that under certain conditions chemicals could organize themselves into complex patterns requiring the coordination of trillions of molecules. And they did this with no instructions. No human organized them. Nor did they have a genetic blueprint that guided their actions. Instead, their own intrinsic self-organizing dynamics directed these complex interactions.

An even simpler way of understanding the creative self-organizing dynamism of matter is to take the perspective of universe and Earth evolution. The deep truth about matter, which neither Descartes nor Newton realized, is that, over the course of four billion years, molten rocks transformed themselves into monarch butterflies, blue herons, and the exalted music of Mozart. Ignorant of this stupendous process, we fell into the fantasy that our role here was to reengineer inert matter.

Our commitment to the control of the natural world has led to the withering of Earth's ecosystems. Life on land and in the oceans is collapsing. The current degradation of Earth is a cataclysmic, biological destruction more catastrophic than anything that has occurred in the past sixty-five million years. Have we come to live in a way so disconnected from Earth's systems of life that this massive destruction is invisible to us? Is it possible that part of the reason we're so out of touch is our inadequate understanding of time?

The traditional, organic sense of time, with its ties to the cycles of nature, was abandoned at the beginning of the modern era. In its place modern humans invented mechanical time. When they enshrined the clock in the city's towers, they disconnected themselves even further from the rhythms of life. Each town learned to organize itself around the position of the clock's hands. There was no longer any real need to glance at the Sun. The machine and its mechanical marking of time slowly became, in effect, the central organizing principle of human life.

This mechanical representation of time enabled an enormous increase in industrial productivity. Humans all over a city could be organized using the same mechanical time. And it was a timepiece that never stopped running. Modern life and work did away with day and night altogether by running nonstop. Everyone from factory workers to investment bankers was tied into networks of machines, some of which supplied materials, others of which shipped away products, all of them tightly meshed together so that the entire process resembled a single complex machine.

Constructing this machine is one of our major achievements. But in doing so, technology became primary, humanity secondary. Disconnected from natural rhythms, humans became slotted into the enveloping industrial patterns. Our systems of housing, transportation, agriculture, and commerce are intertwined and are constructed without significant reference to the patterns of organic life of the enveloping ecosystems. With billions of humans hooked into this vast machine, material production rises but the cost is self-destructive. In addition to the chronic stress, ill-health, and alienation that humans feel inside the machine, there is the unintended consequence of ruining the foundations of every human economy—the ecological processes of life. Each day, ecosystems are deteriorating and irreplaceable species are going extinct, but we modern humans have trained ourselves to focus our minds elsewhere and thus to remain unaware that our industrial civilization is destroying the very conditions for its existence.

But a deeper understanding of time has surfaced within our investigations of nature. We are beginning to understand time as a measure of creative emergence. James Hutton and Charles Lyell in the nineteenth century were some of the first to realize that Earth has emerged over a vast epoch of time involving billions of years. Charles Darwin deepened this awareness through his discoveries of biological evolutionary time. And in the twentieth century, Edwin Hubble and Albert Einstein completed this arc of discovery when they established the development of the universe as a whole over billions of years of creative emergence.

Rather than viewing time as the movement of the hands of

the medieval clock, or the digital display from a vibrating crystal, we can begin to reflect on the way in which time, in a cosmological sense, is the creativity of the universe itself. There was a time for bringing forth hydrogen atoms. There was a time for bringing forth the galaxies. There was a time when Earth became ignited with life. These are indicated not by anything mechanical, but by the deepest processes of the universe itself. There was likewise a time for the universe to bring forth the human species. We live not in any mechanical time, but in this enveloping cosmological time. We live in that time when Earth itself begins its adventure of conscious self-awareness.

ELEVEN

Emerging Earth Community

The challenge of conscious self-awareness is unlike anything that has occurred for millions of years. We are
finding ourselves in the midst of a vast transition. How
are we to respond? For we sense we are in a dark night—we dwell
in unknowing and yet grope forward. The path is still unclear.
With what shall we navigate?

The path is uncertain because our sense of larger purpose and
destiny is clouded. We are seeking patterns that connect us to a
vaster destiny—a vital participation in Earth's unfolding. There is
nothing more mysterious than destiny—of a person, of our species, of our planet, or of the universe itself. But in the modern era

the question was considered unimportant compared with the practical necessities of commerce and trade.

Our puzzlement regarding our destiny is especially poignant since everything else in the universe seems to have a role. The primeval fireball had the work of bringing forth stable matter. The stars had the work of creating the elements. The same is true on Earth. Each species has its unique role to play for the larger community. The phytoplankton in the oceans fill the air with oxygen and thus enable every animal to breathe. That is their great work, to fill each lung with nourishing breath.

But do we humans have such a role? With respect to the universe itself—is there a reason for our existence? Is there a great work required of us?

Throughout the modern period, we have often been dissatisfied with the traditional answers concerning human destiny. Maybe this restlessness reveals something significant in our deep nature. Other species found their biome and settled into it, but nothing has seemed to satisfy us fully. Every place we went we felt we were at home, yet not at home. Some urge carried us forward from place to place.

Perhaps our destiny has something to do with this desire to journey and to experience the depths of things. Perhaps that is why we are here—to drink so deeply of the powers of the universe we become the human form of the universe. Becoming not just nation-state people, but universe people. Becoming a form of human being that is as natural to the universe as the stars or the oceans; knowing how we belong and where we belong so that we enhance the flourishing of the Earth community.

WONDER AND THE STARS

In this process of becoming human we are searching for ongoing guidance. We will need to know what we can rely on. So many of our former certainties are gone now, or are in the process of changing. In order to move into the future we need to know what will be there for us.

First of all, there are the stars. We can count on their presence, their immense fiery light. In the depths of night they are a reassurance that we can find our way. They stun us with their beauty, drawing us into wonder. This sense of wonder is one of our most valuable guides on this ongoing journey into our future as full human beings.

Wonder is a gateway through which the universe floods in and takes up residence within us. Consider the stars. They shine down on Earth for four and a half billion years. Then these new creatures emerged, these humans. What was different about them is that they were amazed every time they beheld the stars. Their amazement inspired works of art and science. Hundreds of thousands of years later, humans discovered that it was these stars that forged the elements of their bodies.

By dwelling in a world of wonder, humans were led to realize that they were children of the stars—something intuited in early myths and uncovered by modern science. They came to understand that everything in the universe then forms a huge interconnected family that we can call "all my relations."

Wonder is not just another emotion; it is rather an opening into the heart of the universe. Wonder is the pathway into what it means to be human, to taste the lusciousness of sun-ripened fruit, to endure the bleak agonies of heartbreak, to exult over the majesty of existence.

The universe's energies penetrate us and awaken us. Through each moment of wonder, no matter how small, we participate in the entrance of primal energies into our lives.

However insignificant we may feel with respect to the age and size of the universe, we are, even so, beings in whom the universe shivers in wonder at itself. By following this wonder we have discovered the ongoing story of the universe, a story that we tell, but a story that is also telling us.

INTIMACY AND THE OCEANS

The oceans too will be our guide as we journey into the future. The ocean is a power that can dissolve things into itself. Even the hardest rocks, given enough time, will become one with the ocean's waves.

With our symbolic consciousness, we are very much like the ocean with its power to pour through boundaries. What we long for is profound intimacy of relationship. Our human imagination brought something radically new to Earth's life: the capacity to experience the world from another's perspective. We call this empathy. What does this mean? In the mammalian world, a mother bear has the capacity to identify with her young cubs and thus devote herself to their well-being. With the emergence of humans, we have arrived at an evolutionary breakthrough for being able to develop compassion, not just for our offspring, but for all beings of every order of existence. With this alone, Earth gave rise to the possibility of an empathetic being who could flow into and become one with the intimate feelings of any being.

Our human destiny is to become the heart of the universe that embraces the whole of the Earth community. We are just a speck in the universe, but we are beings with the capacity to feel comprehensive compassion in the midst of an ocean of intimacy. That is the direction of our becoming more fully human.

CREATIVITY AND FLOURISHING

Finally, along with the stars and the oceans, we can consider what we make with our hands as a way to reflect on human destiny. Our urge to make things, to create things, is certainly as deep as the urge of the Sun to shine and the Earth to spin. Our destiny is woven into the mystery of creativity and time.

For one of the most stunning discoveries in twentieth century cosmology is the deep sense of time that the universe carries. This is not a mechanical time but a cosmological time of universe emergence. In some remarkable way the universe seems to be similar to the unfolding of a giant red oak, where one stage of development leads to the next, as when the galaxies began to form several hundred million years after the birth of the universe. It was not possible for galaxies to emerge earlier or later. If any galaxies started to appear in the first hundred thousand years they would have been torn apart instantly. The energy conditions and patterns of organization necessary for such a magnificent feat were present then and only then. And with the galaxies in a stable form, it was suddenly possible for second- and higher-generation stars with planets to form. These solar systems could not have formed in an earlier era.

It is the same with our moment. We are in the midst of vast destruction, but it is simultaneously a moment of profound creativity. We are involved with building a new era of Earth's life. Our human role is to deepen our consciousness in resonance with the dynamics of the fourteen-billion-year creative event in which we find ourselves. Our challenge now is to construct livable cities and to cultivate healthy foods in ways congruent with Earth's patterns.

Our role is to provide the hands and hearts that will enable the universe's energies to come forth in a new order of well-being. Our destiny is to bring forth a planetary civilization that is both culturally diverse and locally vibrant, a multiform civilization that will enable life and humanity to flourish.

THE ENVELOPING POWERS

Because we know that life is an adventure involving both chaos and order we sometimes want desperately to control things. And whenever our fear grows too strong we become vulnerable to simplistic promises concerning the future. But no one knows what the future holds—all of that is hidden in the darkest night. The future is being created by all of us, and it is a messy and confusing process. What is needed is courage to live in the midst of the ambiguities of this moment without drawing back into fear and a compulsion to control.

Are there guarantees? No, none. But there are reasons for confidence.

When the universe was just quarks and leptons, could anyone have known that it was in the process of bringing forth stars and galaxies? Or later, when Earth emerged, and life existed in the form of tiny jiggling cells, could anyone have seen in them the possibilities of the bluefin tuna or a vast temperate rain forest? We find ourselves inside an amazing drama filled with danger and risk but also stunning creativity. This has happened many times in the past. Two billion years ago, when the atmosphere became so filled with oxygen, all of life was deteriorating. The only way for the life of that time to survive was to burrow deep into the mud at the bottom of the oceans. The future of Earth seemed bleak. And yet, in the midst of that crisis a new kind of cell emerged, one that was not destroyed by oxygen, but was in fact energized by it. Because of this miracle of creativity, life exploded with an exuberance never seen before.

It is in the nature of the universe to move forward between great tensions, between dynamic opposing forces. If the creative energies in the heart of the universe succeeded so brilliantly in the past, we have reason to hope that such creativity will inspire us and guide us into the future. In this way, our own generativity becomes woven into the vibrant communities that constitute the vast symphony of the universe.

FORMATION OF THE UNIVERSE

13.7 billion years ago   The beginning of our observable universe

Particles of matter and light expanding away from a hot origin point

The gravitational, strong nuclear, weak nuclear, and electromagnetic interactions begin shaping the unfolding of the universe

Within minutes, the first nuclei are forming

Within half a million years, the first atoms of hydrogen, helium, and lithium are forming

GALAXIES AND STARS

13 billion years ago   Within half a billion years of the beginning, the first massive stars begin to emerge

Clouds of atoms collapse into the primal galaxies

The first galaxies begin conglomerating into larger disk and elliptical galaxies

12 billion years ago   The universe has formed some 100 billion galaxies, including our Milky Way

8–9 billion years ago   The most rapid star formation in the history of the Milky Way. Most of our Milky Way stars form in this period

Star formation continues to the present and will continue far into the future

Supernova explosions spread elements throughout the galaxies beginning 13 billion years ago and continuing into the future

Our Solar System

4.6 billion years ago	Three supernova explosions trigger star formation in one particular disklike cloud in the Orion arm of the Milky Way Galaxy
4.5 billion years ago	Sun is born
4.45 billion years ago	Earth forms and brings forth an atmosphere, oceans, and continents
3.0 billion years ago	Moon's geological activities are frozen

Life

4.0 billion years ago	First cells emerge
3.9 billion years ago	Photosynthesis
2.3 billion years ago	First Ice Ages
2.0 billion years ago	First cells with nuclei, first multicellular organisms
1.0 billion years ago	Sexual reproduction and heterotrophy

Plants and Animals

Paleozoic Era

Cambrian

542 million years ago	Jellies, sea pens, flat worms
488 million years ago	Cambrian extinctions: 80–90 percent of species eliminated

Ordovician

480 million years ago	Supercontinent Gondwana, South America, Africa, Antarctica, and Madagascar joined as a single land mass
440 million years ago	Ordovician catastrophe

Silurian

425 million years ago	Jawed fishes appear; life moves ashore
415 million years ago	Development of the fin

Devonian

395 million years ago	Insects
380 million years ago	Lungs appear in fish
370 million years ago	Devonian catastrophe; invention of the wood cell by the lycopods; the first trees; vertebrates go ashore; amphibians

Carboniferous

350 million years ago	Land-worthy seeds by the conifers
330 million years ago	Wings by the insects
313 million years ago	Reptiles appear, land-worthy eggs

Permian

256 million years ago	Therapsids, warm-blooded reptiles
245 million years ago	Permian extinctions: 75–95 percent of all species are eliminated

Mesozoic Era

Triassic

235 million years ago	Dinosaurs appear, flowers spread
220 million years ago	Pangaea appears, all continents are joined as a single supercontinent
210 million years ago	First mammals; birth of the Atlantic Ocean; breakup of Pangaea

Jurassic

150 million years ago	Birds

Cretaceous

125 million years ago	Marsupial mammals
114 million years ago	Placental mammals
70 million years ago	Primates emerge
65 million years ago	Cretaceous extinctions

Cenozoic Era

Paleocene

55 million years ago	Rodents, bats, early whales, premonkeys, early horses

Eocene

40 million years ago	Various orders of mammals complete
37 million years ago	Cosmic impact: Eocene catastrophe

Oligocene

36 million years ago	Monkeys
35 million years ago	Early cats and dogs
30 million years ago	First apes
25 million years ago	Whales become largest marine animals of all time; carnivores take to the sea and become seals

Miocene

24 million years ago	Grass spreads across land
20 million years ago	Monkeys and apes split
19 million years ago	Early antelopes
15 million years ago	Cosmic impact: Miocene catastrophe
12 million years ago	Gibbons
11 million years ago	Surge in grazing animals
10 million years ago	Orangutans
9 million years ago	Gorillas
8 million years ago	Modern cats

7 million years ago	Elephants
6 million years ago	Modern dogs

Pliocene

5 million years ago	Chimpanzees, hominids: *Australopithecus afarensis*
4.5 million years ago	Modern camels, bears, and pigs
4.0 million years ago	Baboons
3.7 million years ago	Modern horses
3.5 million years ago	Early cattle
3.3 million years ago	Current Ice Ages begin
2.6 million years ago	First humans: *Homo habilis*
1.8 million years ago	Modern big cats, bison, sheep, wild hogs

Pleistocene

1.5 million years ago	Hunters: *Homo erectus*
1.0 million years ago	Mammalian peak
730,000 years ago	Cosmic impact: Pleistocene catastrophe
700,000 years ago	Brown bears
650,000 years ago	Wolves
500,000 years ago	Llamas
200,000 years ago	Cave bears, goats, modern cattle
150,000 years ago	Woolly mammoths
120,000 years ago	Wildcats
72,000 years ago	Polar bears

THE HUMAN JOURNEY

Paleolithic

2.6 million years ago	African origins with *Homo habilis*, stone tools
1.5 million years ago	*Homo erectus*, hunting
500,000 years ago	Clothing, shelter, fire, hand axes

200,000 years ago	*Homo sapiens*; earliest evidence of human art in the caves of South Africa
100,000 years ago	Ritual burials
40,000 years ago	Entering Australia
35,000 years ago	Entering the Americas

AURIGNACIAN

32,000 years ago	Musical instruments

GRAVETTIAN

20,000 years ago	Spears and bows and arrows

MAGDALENIAN

18,000 years ago	Cave paintings in southern Europe

BECOMING A PLANETARY PRESENCE

NEOLITHIC

12,000 BCE	Dogs tamed
10,700 BCE	Sheep and goats tamed in Middle East
10,600 BCE	Settlements in Middle East; wheat and barley cultivated in Middle East
10,000 BCE	Dogs tamed in North America
9,000 BCE	Settlements in Southeast Asia: rice gardeners; water buffalo, pigs, and chickens tamed; painted pottery culture
8,800 BCE	Cattle tamed in Middle East
8,500 BCE	Settlements in the Americas: cultivation of corn, squash, peppers, and beans; weaving in Middle East

8,000 BCE	Irrigation in Middle East; population of Jericho is 2,000
7,500 BCE	Hassuna culture; millet farmers in North China
7,000 BCE	Catal Huyuk population is 5,000
6,400 BCE	Horses tamed in eastern Europe
5,300 BCE	Pottery in the Andes
5,000 BCE	Early European settlements; gourds, squash, cotton, amaranth, and quinoa in the Andes; camels and donkeys tamed in Middle East; elephants tamed in India
4,500 BCE	Peanuts in the Andes
3,500 BCE	World population is 5–10 million people

CLASSICAL CIVILIZATIONS

3,500 BCE	The wheel and cuneiform writing in Sumer
3,000 BCE	Civilization of the Nile in Egypt
2,800 BCE	Indus Valley civilization on the Indus River
2,100 BCE	Minoan civilization on Crete
2,000 BCE	Megalithic structures in Europe
1,750 BCE	Hammurabi's Code in Babylon
1,700 BCE	Earliest origins of the alphabet in Palestinian region; Aryan-Vedic peoples with Sanskrit language enter India
1,525 BCE	Shang Dynasty in northern China
1,250 BCE	Moses
1,200 BCE	Greek settlements; Exodus of Israel from Egypt, Monotheism
1,100 BCE	Olmec civilization in Meso-America
700 BCE	Homer
628 BCE	Zoroaster
600 BCE	Beginning of Greek philosophy
560 BCE	Confucius in China; Buddha in India

550 BCE	Persian Empire
509 BCE	Founding of Roman Republic
450 BCE	Socrates, Plato, Aristotle
327 BCE	Alexander's invasion of the Indus Valley
260 BCE	India unites under Asoka
221 BCE	China unites in the empire of Qin Shi Huang
150 BCE	Zhang Qian establishes route to Bactria
31 BCE	Roman Empire under Augustus Caesar
4 BCE	Jesus
64 CE	Buddhism in China
100 CE	World population is 300 million
300 CE	Classical Mayan civilization
313 CE	Constantine issues edict of Milan forbidding religious persecution and calls first council of Nicea
410 CE	Fall of Rome
570 CE	Mohammed
650 CE	Muslim civilization
790 CE	Vikings reach North America
800 CE	Carolingian Renaissance in Europe under Charlemagne
900 CE	Toltec civilization
925 CE	Arabic numerals
1000 CE	Islamic science
1088 CE	University of Bologna established
1095 CE	Crusades
1115 CE	Compass invented
1200 CE	Inca civilization
1211 CE	Beginning of the Mongolian Empire under Genghis Khan
1215 CE	Magna Carta limits the king's power in England
1271 CE	Marco Polo begins travels
1320 CE	Aztec civilization

1325 CE	Ibn Battuta begins journeys
1347 CE	Black Death; Asian and European population decline
1433 CE	Cheng Ho voyages to India Ocean, Persian Gulf
1450 CE	Gutenberg Bible printed
1453 CE	Byzantine Empire falls to Turks
1492 CE	Columbus sails to America
1500 CE	World population is 400–500 million
1517 CE	Protestant Reformation begins with Martin Luther
1519 CE	Spanish conquer Aztecs and Incas
1607 CE	English settlement of North America begins at Jamestown

THE MODERN WORLD

1600 CE	British East India Company chartered
1623 CE	Japanese policy of isolation
1721 CE	Peter the Great in Russia
1757 CE	British control over India
1763 CE	European powers divide the colonial world
1776 CE	American Revolution
1789 CE	French Revolution
1815 CE	Napoleon defeated at Waterloo
1833 CE	Slavery abolished across British Empire
1841 CE	Opium War settled, with China establishing live trading ports
1848 CE	Revolutions across Europe
1854 CE	Matthew Perry forces Japan open to Western trade
1867 CE	Karl Marx publishes *Das Kapital*
1884 CE	European powers divide Africa into colonies
1914 CE	World War I

1917 CE	Communism takes control in Russia
1918 CE	Influenza pandemic
1919 CE	League of Nations
1933 CE	Great Depression
1939 CE	World War II
1944 CE	Breton Woods conference establishes monetary system
1945 CE	First atomic bomb exploded over Hiroshima; United Nations Charter
1947 CE	Partition of India
1948 CE	Mahatma Gandhi is assassinated
	Founding of Israel
1949 CE	Mao Zedong controls China
1957 CE	Sputnik launched, first artificial satellite
1968 CE	Student riots in France, United States, Japan
1969 CE	Humans set foot on the Moon
1970 CE	Earth Day inaugurated
1972 CE	First picture of Earth from space, called the Blue Marble
1979 CE	Iranian revolution ignites Islamic resurgence
1982 CE	World Charter for Nature
1989 CE	Berlin Wall falls and Cold War ends
1990 CE	Nelson Mandela is freed and apartheid ends in South Africa
1991 CE	Dissolution of Soviet Union; Internet initiated
1992 CE	UN Conference on Environment and Development in Rio de Janeiro; Framework Convention of Climate Change; Convention on Biological Diversity
1995 CE	World Social Summit in Copenhagen; UN Women's Summit in Beijing
1996 CE	UN Human Settlements Summit in Istanbul; World Food Summit in Rome
1999 CE	World population reaches 6 billion

2000 CE	Earth Charter
2002 CE	World Summit on Sustainable Development in Johannesburg
2005 CE	United Nations Alliance of Civilizations formed
2007 CE	United Nations Declaration on the Rights of Indigenous Peoples
2010 CE	Universal Declaration of the Rights of Mother Nature
	Earth Charter + 10 conference in Ahmedabad, India

THE SCIENTIFIC REVOLUTION

1543 CE	Nicolaus Copernicus formulates a heliocentric universe
1609 CE	Johannes Kepler discovers the elliptical movement of the planets around the Sun
1609 CE	Galileo Galilei establishes empirical mode of observation by effectively using precise measurements in his observations of natural phenomena
1620 CE	Francis Bacon promotes a pragmatic orientation of modern science
1637 CE	René Descartes establishes mathematic mode of dealing with the natural world and divides the physical world and mind into two entirely different realms
1687 CE	Isaac Newton explains the modern view of the universe
1749 CE	Georges-Louis Buffon rethinks the age of the Earth
1750 CE	Carolus Linnaeus provides the modern system of taxonomic classification of life

1755 CE	Immanuel Kant proposes a theory of the formation of celestial bodies and the solar system
1795 CE	James Hutton discovers that the geological formation of the Earth and of life can be traced back in time
1809 CE	Jean-Baptiste Lamarck traces the evolutionary sequence from lower forms to higher forms of life
1827 CE	Georges Cuvier sets the basis for the classification of animals
1830 CE	Charles Lyell describes the structure of the Earth
1859 CE	Charles Darwin publishes his theory of natural selection and alters our understanding of the development of life
1866 CE	Gregor Mendel publishes paper on plant hybridization
1905 CE	Albert Einstein alters our basic understanding of time, space, motion, matter, and energy
1912 CE	Alfred Wegener proposes the theory of continental drift
1927 CE	Werner Heisenberg changes our perception of knowledge at the atomic level
1929 CE	Edwin Hubble provides evidence that we live in an expanding universe
1950 CE	Hans Albrecht Bethe describes how stars evolve
1953 CE	James Watson and Francis Crick propose double helix structure of DNA
1962 CE	Rachel Carson exposes the effects of modern pesticides on the natural world
1965 CE	Robert Wilson and Arno Penzias find evidence of the origin of the universe

1969 CE	Neil Armstrong is first human to set foot on the Moon
1972 CE	Niles Eldredge and Steven Jay Gould propose theory of punctuated equilibria
1977 CE	Ilya Prigogine wins Nobel Prize for work on self-organizing dynamics
1984 CE	Theory of cold dark matter proposed
1998 CE	Accelerating expansion of the universe proposed
2003 CE	Human genome project
2010 CE	Since 1990s thousands of extrasolar planets have been discovered
	Discovery of a 13.1-billion-year-old galaxy

NOTES

ONE   Beginning of the Universe

1. Freeman J. Dyson, *Disturbing the Universe* (New York: Harper and Row, 1979), 250.

TWO   Galaxies Forming

1. Ben Zuckerman and Matthew Malkan, eds., *The Origin and Evolution of the Universe* (Sudbury, MA: Jones and Bartlett, 1996), 33.

FOUR   Birth of the Solar System

1. Cheng I, *A Source Book in Chinese Philosophy,* trans. Wing-tsit Chan (Princeton: Princeton University Press, 1963), 298, 553.

SIX   Living and Dying

1. L. V. Salvini-Plawen and E. Mayr in *Evolutionary Biology*, ed. M. K. Hecht, W. C. Steere, and B. Wallace (New York: Plenum, 1977), quoted in John D. Barrow and Frank Tipler, *The Anthropic Cosmological Principle* (New York: Oxford University Press, 1986), 132.

SEVEN   The Passion of Animals

1. Scott D. Sampson, *Dinosaur Odyssey: Fossil Threads in the Web of Life* (Berkeley: University of California Press, 2009), 172.

EIGHT   The Origin of the Human

1. Nicholas Wade, *Before the Dawn* (New York: Penguin, 2006), 75.
2. Christine Kenneally, "Freedom from Selection Lets Genes Get Creative," *New Scientist,* vol. 27 (Sept. 2008): 40–43.

3. Wm. Theodore de Bary and Irene Bloom eds., *Sources of Chinese Tradi-tion,* vol. 1 (New York: Columbia University Press, 1999), 683.

N I N E   Becoming a Planetary Presence

1. The term *Anthropocene* was coined by Nobel Prize–winning at-mospheric chemist Paul Crutzen. He published an article first using the term in the 2000 newsletter of the International Geosphere-Biosphere Programme (IGBP), no. 41.

ONE   Beginning of the Universe

Ambjorn, Jan, Jerzy Jurkiewicz, and Renate Loll. "The Self-Organizing Quantum Universe." *Scientific American.* July, 2008.

Attard, Phil. "The Second Law of Nonequilibrium Thermodynamics." *Advances in Chemical Physics.* 140/1, 2008.

Barrow, John. *The Origin of the Universe.* New York: Basic Books, 2001.

Barrow, John, Paul C. W. Davies, and Charles L. Harper Jr., eds. *Science and Ultimate Reality: Quantum Theory, Cosmology, and Complexity.* Cambridge: Cambridge University Press, 2004.

Carr, Bernard, ed. *Universe or Multiverse?* Cambridge: Cambridge University Press, 2007.

Chaisson, Eric. *The Epic of Evolution: Seven Ages of the Cosmos.* New York: Columbia University Press, 2005.

Greene, Brian. *The Elegant Universe.* New York: Vintage Books, 1996.

Hakim, Joy. *The Story of Science: Einstein Adds a New Dimension.* Washington, DC: Smithsonian Books, 2007.

Leslie, John, ed. *Physical Cosmology and Philosophy.* New York: Macmillan, 1990.

North, John. *The Norton History of Astronomy and Cosmology.* New York: W. W. Norton and Company, 1995.

Primack, Joel, and Nancy Ellen Abrams. *View from the Center of the Universe: Discovering Our Extraordinary Place in the Cosmos.* New York: Riverhead-Penguin, 2007.

Pylkkanen, Paavo. *Mind, Matter, and the Implicate Order.* Berlin: Springer, 2007.

Rees, Martin. *Just Six Numbers: The Deep Forces That Shape the Universe.* New York: Basic Books, 1997.

———. *Our Cosmic Habitat.* Princeton: Princeton University Press, 2001.

Reeves, Hubert. *Hour of Our Delight: Cosmic Evolution, Order, and Complexity.* San Francisco: Freeman, 1991.

Rickles, Dean, Steven French, and Juha Saatsi, eds. *The Structural Foundations of Quantum Gravity.* Oxford: Oxford University Press, 2006.

Schilling, Govert. *Evolving Cosmos.* Cambridge: Cambridge University Press, 2004.

Silk, Joseph. *Horizons of Cosmology.* West Conshohocken, PA: Templeton Press, 2009.

Smolin, Lee. *The Life of the Cosmos.* New York: Oxford University Press, 1999.

Teerikorpi, Pekka, Mauri Valtonen, K. Lehto, Harry Lehto, Gene Byrd, and Arthur Chernin. *The Evolving Universe and the Origin of Life: The Search for Our Cosmic Roots.* Berlin: Springer, 2009.

Vaas, Rudiger, ed. *Beyond the Big Bang: Competing Scenarios for an Eternal Universe.* Berlin: Springer, 2011.

Verdal, Vlatko. *Decoding Reality: The Universe as Quantum Information.* Oxford: Oxford University Press, 2010.

T W O   Galaxies Forming

Aschwanden, Markus. *Self-Organized Criticality in Astrophysics.* Berlin: Springer, 2011.

Baryshev, Yurji, and Pekka Teerikorpi. *Discovery of Cosmic Fractals.* Singapore: World Scientific, 2002.

Chaisson, Eric. *Cosmic Evolution: The Rise of Complexity in Nature.* Cambridge, MA: Harvard University Press, 2002.

Davies, Paul. *The Goldilocks Enigma: Why Is the Universe Just Right for Life.* London: Allen Lane-Penguin, 2006.

Dyson, Freeman. "Time Without End: Physics and Biology in an Open Universe." *Reviews of Modern Physics* 51/3 (1979): 447–460.

Freedman, Roger A., and William Kaufman. *Universe.* 8th ed. New York: W. H. Freeman, Macmillan, 2007.

Gabrielli, Andrea, Francesco Sylos Labini, Michael Joyce, and Luciano Pietronero. *Statistical Physics for Cosmic Structures.* New York: Springer, 2005.

Laughlin, Robert B. *A Different Universe: Reinventing Physics from the Bottom Down.* New York: Basic Books: 2006.

Liddle, Andrew, and Jon Loveday. *The Oxford Companion to Cosmology.* Oxford: Oxford University Press, 2009.

Longair, Malcolm S. *Galaxy Formation. Astronomy and Astrophysics Library.* 2d ed. New York: Springer, 2008.

Mo, Houjun, Simon White, and Frank van den Bosch. *Galaxy Formation and Evolution.* Cambridge: Cambridge University Press, 2010.

Morrison, Philip, and Morrison, Phylis. *Powers of Ten: A Book About the Relative Size of Things in the Universe and the Effect of Adding Another Zero.* Redding, CT: Scientific American Library, 1982.

Pagel, Bernard. *Nucleosynthesis and Chemical Evolution of Galaxies.* 2d ed. Cambridge: Cambridge University Press, 2009.

Sagan, Carl. *Cosmos.* New York: Random House, 1980.

Sparke, Linda, and John Gallagher. *Galaxies in the Universe: An Introduction.* Cambridge: Cambridge University Press, 2007.

Tyson, Neil deGrasse, and Donald Goldsmith. *Origins: Fourteen Billion Years of Cosmic Evolution.* New York: W. W. Norton, 2004.

THREE   The Emanating Brilliance of Stars

Ankay, Askin, Oktay H. Guseinov, and Efe Yazgan. *Neutron Stars, Super-novae, and Supernova Remnants.* Hauppauge, NY: Nova Science, 2007.

Duncan, Todd, and Craig Tyler. *Your Cosmic Context: An Introduction to Modern Cosmology.* San Francisco: Benjamin Cummings-Pearson, 2008.

Eales, Stephen. *Origins: How the Planets, Stars, Galaxies, and the Universe Began.* Berlin: Springer, 2006.

Kippenhahn, Rudolf. *100 Billion Suns: The Birth, Life, and Death of the Stars.* New York: Basic Books, 1983.

LeBlanc, Francis. *An Introduction to Stellar Astrophysics.* New York: Wiley, 2010.

Murdin, Paul. *End in Fire: The Supernova in the Large Magellanic Cloud.* New York: Cambridge University Press, 1990.

Ollivier, Marc, Thérèse Encrenaz, Françoise Roques, Franck Selsis, and Fabienne Casoli. *Planetary Systems: Detection, Formation, and Habitability of Extrasolar Planets.* Berlin: Springer, 2009.

Prialnik, Dina. *An Introduction to the Theory of Stellar Structure and Evolution.* New York: Cambridge University Press, 2009.

Ryan, Sean, and Andrew Norton. *Stellar Evolution and Nucleosynthesis.* Cambridge: Cambridge University Press, 2010.

FOUR   Birth of the Solar System

Casoli, Fabienne, and Therese Encrenaz. *The New Worlds: Extrasolar Planets.* Berlin: Springer, 2010.

Chela-Flores, Julian. *A Second Genesis: Stepping-Stones Towards the Intelligibility of Nature.* Singapore: World Scientific, 2009.

Condie, Kent. *Earth as an Evolving Planetary System.* Boston: Academic Press, 2004.

Dobretsov, Nikolay, Nikolay Kolchanov, Alexey Rozanov, and Georgy Zavarzin, eds. *Biosphere Origin and Evolution.* New York: Springer, 2008.

Ehrenfreund, Pascale, W. M. Irvine, T. Owen, Luann Becker, Jen Blank, J. R. Brucato, and Luigi Colangeli. *Astrobiology: Future Perspectives.* New York: Springer, 2004.

Hazen, Robert, Dominic Papineau, Wouter Bleeker, Robert T. Downs, John M. Ferry, Timothy J. McCoy, Dimitri A. Sverjensky, and Hexiong Yang. "Mineral Evolution." *American Mineralogist.* 93/1693, 2008.

Hergarten, Stefan. *Self-Organized Criticality in Earth Systems.* Berlin: Springer, 2002.

Lin, Douglas. "The Genesis of Planets." *Scientific American.* May, 2008.

Livio, Mario, N. Reid, and W. Sparks, eds. *Astrophysics of Life.* Cambridge: Cambridge University Press, 2005.

Lyell, Charles. *Principles of Geology.* First published 1830–1833.

Margulis, Lynn, Clifford Matthews, and Aaron Haselton, eds. *Environmental Evolution.* Cambridge, MA: MIT Press, 2000.

Ord, Alison, Giles W. Hunt, and Bruce E. Hobbs, eds. "Patterns in Our Planet: Defining New Concepts for the Applications of Multi-scale Non-equilibrium Thermodynamics to Earth-system Science." *Philosophical Transactions of the Royal Society A.* 368/3, 2010.

Poole, Robert. *Earthrise.* New Haven: Yale University Press, 2008.

Pudritz, Ralph, Paul Higgs, and Jonathon Stone, eds. *Planetary Systems and the Origins of Life.* Cambridge: Cambridge University Press, 2007.

Stewart, Iain, and John Lynch. *Earth: The Biography.* Washington, DC: National Geographic, 2007.

Tsonis, Anastasios, and James Elsner, eds. *Nonlinear Dynamics in Geosciences.* Berlin: Springer, 2007.

FIVE   Life's Emergence

Abbott, Derek, Paul C. W. Davies, and Arun K. Pati, eds. *Quantum Aspects of Life.* Singapore: World Scientific, 2008.

Alvarez, Walter. *T. Rex and the Crater of Doom.* Princeton: Princeton University Press, 1997.

Baltimore, David, Renato Dulbecco, Francois Jacob, and Rita Levi-Montalcini, eds. *Frontiers of Life.* 4 vol. San Diego: Academic Press, 2002.

Bateson, Gregory. *Mind and Nature: A Necessary Unity.* New York: Bantam Books, 1988.

Bedau, Mark, and Carol Cleland, eds. *The Nature of Life: Classical and Contemporary Perspectives from Philosophy and Science.* Cambridge: Cambridge University Press, 2010.

Crofts, Antony. "Life, Information, Entropy, and Time." *Complexity* 13/1 (2007): 14–50.

Deamer, David, and Jack Szostak, eds. *The Origins of Life.* Cold Spring Harbor, NY: Cold Spring Harbor Laboratory Press, 2010.

De Duve, Christian. *Singularities: Landmarks on the Pathways of Life.* New York: Columbia University Press, 2005.

Fisher, George W., Grace S. Brush, and Philip D. Curtin. *Discovering the Chesapeake: The History of an Ecosystem.* Baltimore: Johns Hopkins University Press, 2001.

Goodenough, Ursula. *The Sacred Depths of Nature.* New York: Oxford University Press, 2000.

Harold, Franklin. *The Way of the Cell.* Oxford: Oxford University Press, 2001.

Hazen, Robert. *Genesis: The Scientific Quest for Life's Origin.* Washington, DC: Joseph Henry Press, 2005.

Hickey, Leo. *The Forest Primeval: The Geologic History of Wood and Petrified Forests.* Yale University Publications in Anthropology. New Haven: Yale Peabody Museum, 2010.

Knoll, Andrew. *Life on a Young Planet: The First Three Billion Years of Evolution on Earth.* Princeton: Princeton University Press, 2003.

Liebes, Sidney, Elisabet Sahtouris, and Brian Swimme. *A Walk Through Time: From Star Dust to Us.* New York: John Wiley, 1998.

Lovejoy, Thomas, John Browne, and Chris Patten. *Respect for the Earth: Sustainable Development: Reith Lecture.* London: Profile Books, 2000.

Luisi, Pier Luigi. *The Emergence of Life: From Chemical Origins to Synthetic Biology.* Cambridge: Cambridge University Press, 2010.

Meinesz, Alexandre. *How Life Began: Evolution's Three Geneses.* Chicago: University of Chicago, 2008.

Noble, Denis. *The Music of Life.* Oxford: Oxford University Press, 2006.

Rasmussen, Steen, Mark A. Bedau, Liaohai Chen, David Deamer, David C. Krakauer, Norman H. Packard, and Peter F. Stadler, eds. *Protocells: Bridging Nonliving and Living Matter.* Cambridge, MA: MIT Press, 2009.

Raven, Peter, and Linda R. Berg. *Environment.* 7th ed. New York: Wiley, 2009.

Rhodes, Frank, Richard O. Stone, and Bruce D. Malamud, eds. *Language of the Earth.* Malden, MA: Blackwell, 2008.

Russell, Dale. *Islands in the Cosmos: The Evolution of Life on Land.* Bloomington, IN: Indiana University Press, 2009.

Sampson, Scott. *Dinosaur Odyssey: Fossil Threads in the Web of Life.* Berkeley: University of California Press, 2009.

Sapp, Jan. *The New Foundations of Evolution.* Oxford: Oxford University Press, 2009.

Schneider, Stephen, James R. Miller, Eileen Crist, and Penelope J. Boston, eds. *Scientists Debate Gaia.* Cambridge, MA: MIT Press, 2004.

Seckbach, Joseph, ed. *Origins: Genesis, Evolution and Diversity of Life.* Dordrecht: Kluwer Academic, 2004.

Woese, Carl. "A New Biology for a New Century." *Microbiology and Molecular Biology Reviews* 68/2 (2004): 173–186.

Zaikowski, Lori, Jon Friedrich, and S. Russell Seidel, eds. *Chemical Evolution II: From the Origins of Life to Modern Society.* Washington, DC: American Chemical Society, 2009.

Zewail, Ahmed, ed. *Physical Biology: From Atoms to Medicine.* London: Imperial College Press, 2008.

S I X   Living and Dying

Ackerman, Diane. *A Natural History of the Senses.* New York: Vintage, 1991.

Bak, Per. *How Nature Works: The Science of Self-Organized Criticality.* New York: Springer, 1999.

Barabasi, Albert-Laszlo. *Linked: The New Science of Networks.* Cambridge, MA: Perseus Books, 2002.

Barash, David P. *Natural Selections: Selfish Altruists, Honest Liars, and Other Realities of Evolution.* New York: Bellevue Literary Press, 2007.

Barberousse, Anouk, Michel Morange, and Thomas Pradeu, eds. *Mapping the Future of Biology.* Berlin: Springer, 2009.

Bar Yam, Yaneer. *Dynamics of Complex Systems.* Reading, MA: Addison-Wesley, 1997.

Blumberg, Mark, John Freeman, and Scott Robinson, eds. *Oxford Handbook of Developmental Behavioral Neuroscience.* Oxford: Oxford University Press, 2010.

Callebaut, Werner, and Diego Rasskin-Gutman, eds. *Modularity: Understanding the Development and Evolution of Natural Complex Systems.* Cambridge, MA: MIT Press, 2005.

Camazine, Scott, Jean-Louis Deneubourg, Nigel R. Franks, James Sneyd, Guy Theraulaz, and Eric Bonabeau, eds. *Self-Organization in Biological Systems.* Princeton: Princeton University Press, 2001.

Caporale, Lynn, ed. *The Implicit Genome.* Oxford: Oxford University Press, 2006.

Carroll, Sean B. *Endless Forms Most Beautiful: The New Science of Evo Devo.* New York: Norton, 2005.

Conway Morris, Simon, ed. *The Deep Structure of Biology: Is Convergence Sufficiently Ubiquitous to Give a Directional Signal?* West Conshohocken, PA: Templeton Foundation Press, 2008.

Darwin, Charles. *Origin of Species.* New York: Signet Classics/Penguin Books, 2003. First published 1859.

Jablonka, Eva, and Marion Lamb. *Evolution in Four Dimensions: Genetic,*

Epigenetic, Behavioral, and Symbolic Variation in the History of Life. Cambridge, MA: MIT Press, 2006.

Kirschner, Mark, and John Gerhart. *The Plausibility of Life.* New Haven: Yale University Press, 2005.

Laubichler, Manfred, and Jane Maienschein, eds. *From Embryology to Evo-Devo.* Cambridge, MA: MIT Press, 2007.

Margulis, Lynn. *Symbiotic Planet: A New Look at Evolution.* Reading, MA: Perseus Books, 2000.

Meyers, Robert, editor-in-chief. *Encyclopedia of Complexity and Systems Science.* Berlin: Springer, 2009.

Mitchell, Sandra. *Biological Complexity and Integrative Pluralism.* Cambridge: Cambridge University Press, 2003.

Morris, Simon Conway. *Life's Solution: Inevitable Humans in a Lonely Universe.* New York: Cambridge University Press, 2004.

Neumann-Held, Eva, and Christoph Rehmann-Sutter, eds. *Genes in Development: Re-reading the Molecular Paradigm.* Durham, NC: Duke University Press, 2006.

Nowak, Martin. *Evolutionary Dynamics.* Cambridge, MA: Harvard University Press, 2006.

Nuland, Sherwin. *The Wisdom of the Body.* New York: Alfred A. Knopf, 1997.

Pigliucci, Massimo, and Gerd Muller, eds. *Evolution—the Extended Synthesis.* Cambridge, MA: MIT Press, 2010.

Reid, Robert G. B. *Biological Emergences.* Cambridge, MA: MIT Press, 2007.

Richerson, Peter. *Not by Genes Alone: How Culture Transformed Human Evolution.* Chicago: University of Chicago Press, 2006.

Ruse, Michael, and Joseph Travis, eds. *Evolution: The First Four Billion Years.* Cambridge, MA: Harvard University Press, 2009.

Sansom, Roger, and Robert Brandon, eds. *Integrating Evolution and Development.* Cambridge, MA: MIT Press, 2007.

Smith, John Maynard, and Eörs Szathmáry. *The Major Transitions in Evolution.* New York: Oxford University Press, 1998.

Sole, Ricard, and Jordi Bascompte. *Self-Organization in Complex Ecosystems.* Princeton: Princeton University Press, 2006.

Stewart, Ian. "Self-organization in Evolution." *Philosophical Transactions of the Royal Society of London,* special issue on Self-Organization: The Quest for the Origin and Evolution of Structure (June 15, 2003): 1101–1123.

Strogatz, Steven H. *Sync: How Order Emerges from Chaos in the Universe, Nature, and Daily Life.* New York: Hyperion, 2004.

Weber, Bruce, and David Depew, eds. *Evolution and Learning.* Cambridge, MA: MIT Press, 2003.

Witzany, Gunther. *Biocommunication and Natural Genome Editing.* Dordrecht: Springer, 2010.

Wrangham, Richard. *Catching Fire: How Cooking Made Us Human.* New York: Basic Books, 2009.

Zimmer, Carl. *The Tangled Bank: An Introduction to Evolution.* Greenwood Village, CO: Roberts and Co. Publishers, 2009.

S E V E N   The Passion of Animals

Bekoff, Marc. *Animal Passions and Beastly Virtues: Reflections on Redecorating Nature.* Philadelphia: Temple University Press, 2005.

——, ed. *The Encyclopedia of Animal Behavior.* Santa Barbara, CA: Greenwood, 2004.

Birch, Charles, and John B. Cobb. *The Liberation of Life: From the Cell to the Community.* New York: Cambridge University Press, 1981.

Cheney, Dorothy, and Robert Seyfarth. *Baboon Metaphysics: The Evolution of a Social Mind.* Chicago: University of Chicago Press, 2008.

Cobb, John, and David Griffin, eds. *Mind in Nature: Essays on the Interface of Science and Philosophy.* Washington, DC: University Press of America, 1977.

Corning, Peter. *Nature's Magic: Synergy in Evolution and the Fate of Humankind.* New York: Cambridge University Press, 2003.

Croft, Darren, Richard James, and Jens Krause. *Exploring Animal Social Networks.* Princeton: Princeton University Press, 2008.

Goettner-Abendroth, Heide, ed. *Societies of Peace: Matriarchies Past, Present and Future.* Toronto: Inanna Publications, 2009.

Hemelrijk, Charlotte, ed. *Self-Organization and Evolution of Social Systems.* Cambridge: Cambridge University Press, 2005.

Hrdy, Sarah Blaffer. *Mothers and Others: The Evolutionary Origins of Mutual Understanding.* Cambridge, MA: Harvard University Press, 2009.

Hurley, Susan, and Matthew Nudds, eds. *Rational Animals?* Oxford: Oxford University Press, 2006.

Laland, Kevin, and Bennett Galef, eds. *The Question of Animal Culture.* Cambridge, MA: Harvard University Press, 2009.

Levine, George. *Darwin Loves You: Natural Selection and the Re-Enchantment of the World.* Princeton: Princeton University Press, 2006.

Reznikova, Zhanna. *Animal Intelligence: From Individual to Social Cognition.* Cambridge: Cambridge University Press, 2007.

Robb, Christina. *This Changes Everything: The Relation Revolution in Psychology.* New York: Farrar, Straus and Giroux, 2006.

Rogers, Lesley J. *Minds of Their Own: Thinking and Awareness in Animals.* Boulder, CO: Westview Press, 1998.

Roughgarden, Joan. *The Genial Gene: Deconstructing Darwinian Selfishness.* Berkeley: University of California Press, 2009.

Smuts, Barbara. *Sex and Friendship in Baboons.* New York: Aldine Publishing Co., 1985 (republished 2009).

Sober, Elliott, and David Sloan Wilson. *Unto Others: The Evolution and Psychology of Unselfish Behavior.* Cambridge, MA: Harvard University Press, 1998.

Stanley, Steven M. *Earth and Life Through Time.* 2d ed. New York: Freeman, 1989.

Tomasello, Michael. *Why We Cooperate.* Cambridge, MA: MIT Press, 2009.

Waldau, Paul, and Kimberly Patton, eds. *A Communion of Subjects: Animals in Religion, Science, and Ethics.* New York: Columbia University Press, 2006.

Weiss, Kenneth, and Anne Buchanan. *The Mermaid's Tale: Four Billion Years of Cooperation in the Making of Living Things.* Cambridge, MA: Harvard University Press, 2009.

Wilson, David Sloan. *Darwin's Cathedral.* Chicago: University of Chicago Press, 2002.

EIGHT   The Origin of the Human

Barbieri, Marcello, ed. *Introduction to Biosemiotics: The New Biological Synthesis.* Berlin: Springer, 2007.

Bentley, Alexander, and Herbert D. G. Maschner, eds. *Complex Systems and Archaeology: Empirical and Theoretical Applications.* Salt Lake City, UT: University of Utah Press, 2003.

Boyden, Stephen. *The Biology of Civilization.* Sydney: University of New South Wales Press, 2004.

Bromade, Timothy, and Friedemann Shrenck, eds. *African Biogeography, Climate Change, and Human Evolution.* New York: Oxford University Press, 1999.

Campbell, Bernard G. *Humankind Emerging.* 5th ed. Boston: Scott, Foresman, 1988.

Deacon, Terrence W. *The Symbolic Species: The Co-evolution of Language and the Brain.* New York: W. W. Norton and Company, 1998.

De Waal, Frans. *Our Inner Ape: A Leading Primatologist Explains Why We Are Who We Are.* New York: Riverhead-Penguin, 2006.

Diamond, Jared. *Guns, Germs, and Steel.* New York: Norton, 1998.

Donald, Merlin. *A Mind So Rare.* New York: Norton, 2001.

Eliade, Mircea. *Shamanism: Archaic Techniques of Ecstasy.* Princeton, NJ: Princeton University Press, 1974.

Ellis, Nick, and Diane Larsen-Freeman, eds. *Language as a Complex Adaptive System.* New York: Wiley/Blackwell, 2010.

Falk, Dean. *Finding Our Tongues: Mothers, Infants, and the Origins of Language.* New York: Basic Books, 2009.

Fernandez, Eliseo. "Taking the Relational Turn: Biosemiotics and Some New Trends in Biology." *Biosemiotics* 3/2 (2010): 147–156.

Gontier, Nathalie, Jean Paul van Bendegem, and Diederik Aerts, eds. *Evolutionary Epistemology, Language and Culture: A Non-Adaptationist, Systems Theoretical Approach.* Dordrecht: Springer, 2006.

Grim, John A. *The Shaman: Patterns of Siberian and Ojibway Healing.* Norman, OK: University of Oklahoma Press, 1984.

Hoffmeyer, Jesper. *Biosemiotics: An Examination into the Signs of Life and the Life of Signs.* Chicago: University of Chicago Press, 2009.

Hornborg, Alf, and Carole Crumley, eds. *The World System and the Earth System.* Walnut Creek, CA: Left Coast Press, 2007.

Johanson, Donald, and Kate Wong. *Lucy's Legacy: The Quest for Human Origins.* New York: Harmony Books, 2009.

Jolly, Alison. *Lucy's Legacy: Sex and Intelligence in Human Evolution.* Cambridge, MA: Harvard University Press, 1999.

Konner, Melvin. *The Evolution of Childhood.* Cambridge, MA: Harvard University Press, 2010.

Mithen, Steven. *The Singing Neanderthals: The Origins of Music, Language, Mind and Body.* Cambridge, MA: Harvard University Press, 2006.

Mufwene, Salikoko. *The Ecology of Language Evolution.* Cambridge: Cambridge University Press, 2001.

Namhee, Lee, Lisa Mikesell, Anna Dina L. Joaquin, Andrea W. Mates, and John H. Schumann, eds. *The Interactional Instinct: The Evolution and Acquisition of Language.* Oxford: Oxford University Press, 2009.

Tatersall, Ian. *Becoming Human: Evolution and Human Uniqueness.* New York: Harcourt Brace, 1998.

Tomasello, Michael. *The Cultural Origins of Human Cognition.* Cambridge, MA: Harvard University Press, 2001.

Tononi, Giulio. "Consciousness as Integrated Information." *Biological Bulletin* 215/3 (2008): 216–242.

Wautischer, Helmut, ed. *Ontology of Consciousness: Percipient Action.* Cambridge, MA: MIT Press, 2008.

Wimberley, Edward. *Nested Ecology: The Place of Humans in the Ecological Hierarchy.* Foreword by John Haught. Baltimore: Johns Hopkins University Press, 2009.

NINE   Becoming a Planetary Presence

Berry, Wendell. *The Unsettling of America: Culture and Agriculture.* San Francisco: Sierra Club Books, 1977.

Botkin, Daniel. *Discordant Harmonies: A New Ecology for the Twenty-First Century.* New York: Oxford University Press, 1990.

Bull, Hedley, and Adam Watson, eds. *The Expansion of International Society.* Oxford: Clarendon Press, 1984.

Callicott, J. Baird. *Earth's Insights: A Multicultural Survey of Ecological Ethics from the Mediterranean Basin to the Australian Outback.* Berkeley: University of California Press, 1997.

Carrasco, David. *Quetzalcoatl and the Irony of Empire: Myths and Prophecies in the Aztec Tradition.* Chicago: University of Chicago Press, 1982.

Chapple, Christopher Key, ed. *Jainism and Ecology: Nonviolence in the Web of Life.* Religions of the World and Ecology Series, edited by Mary Evelyn Tucker and John Grim. Cambridge, MA: Harvard Center for the Study of World Religions, 2002.

Chapple, Christopher Key, and Mary Evelyn Tucker, eds. *Hinduism and Ecology: The Intersection of Earth, Sky, and Water.* Religions of the World and Ecology Series, edited by Mary Evelyn Tucker and John Grim. Cambridge, MA: Harvard Center for the Study of World Religions, 2000.

Christian, David. *Maps of Time: An Introduction to Big History.* Berkeley: University of California Press, 2004.

Costanza, Robert, Lisa J. Graumlich, and Will Steffen, eds. *Sustainability or Collapse? An Integrated History and Future of People on Earth.* Cambridge, MA: MIT Press, 2007.

de Bary, William Theodore, Donald Keene, George Tanabe, and Paul Varley, eds. *Sources of Japanese Tradition.* New York: Columbia University Press, 2002.

de Bary, William Theodore, Richard Lufrano, Irene Bloom, and Joseph Adler, eds. *Sources of Chinese Tradition.* 2 vol. New York: Columbia University Press, 1999–2000.

Deming, Alison Hawthorne. *Writing the Sacred into the Real.* Minneapolis: Milkweed, 2001.

Diamond, Jared. *Collapse: How Societies Choose to Fail or Succeed.* New York: Viking, 2005.

Ehrlich, Paul. *Human Natures: Genes, Cultures, and the Human Prospect.* Washington, DC: Island Press, 2000.

Eldredge, Niles, ed. *Life on Earth: An Encyclopedia of Biodiversity, Ecology, and Evolution.* Santa Barbara, CA: ABC-CLIO, 2002.

Embree, Ainsley, William Theodore de Bary, and Stephen N. Hay, eds. *Sources of Indian Tradition.* New York: Columbia University Press, 1988.

Escobar, Arturo. *Territories of Difference: Place, Movement, Life, Redes.* Durham, NC: Duke University Press, 2008.

Foltz, Richard C., Frederick M. Denny, and Azizan Baharuddin, eds. *Islam and Ecology: A Bestowed Trust.* Religions of the World and Ecology Series, edited by Mary Evelyn Tucker and John Grim. Cambridge, MA: Harvard Center for the Study of World Religions, 2003.

Girardot, Norman J., James Miller, and Liu Xiaogan, eds. *Daoism and*

Ecology: Ways Within a Cosmic Landscape. Religions of the World and Ecology Series, edited by Mary Evelyn Tucker and John Grim. Cambridge, MA: Harvard Center for the Study of World Religions, 2001.

Goodstein, David. *Out of Gas: The End of the Age of Oil.* New York: Norton, 2004.

Grim, John A., ed. *Indigenous Traditions and Ecology: The Interbeing of Cosmology and Community.* Religions of the World and Ecology Series, edited by Mary Evelyn Tucker and John Grim. Cambridge, MA: Harvard Center for the Study of World Religions, 2001.

Hamilton, Clive. *Growth Fetish.* London: Pluto Press, 2003.

Hessel, Dieter T., and Rosemary Radford Ruether, eds. *Christianity and Ecology: Seeking the Well-being of Earth and Humans.* Religions of the World and Ecology Series, edited by Mary Evelyn Tucker and John Grim. Cambridge, MA: Harvard Center for the Study of World Religions, 2000.

Karan, Pradyumna. *The Non-Western World.* New York: Routledge, 2004.

Kebede, Messay. *Africa's Quest for a Philosophy of Decolonization.* Amsterdam: Rodopi, 2004.

Kellert, Stephen, and Timothy Farnham, eds. *The Good in Nature and Humanity: Connecting Science, Religion, and Spirituality with the Natural World.* Washington, DC: Island Press, 2002.

Khaldoun, Ibn. *The Muqaddimah: An Introduction to History.* 3 vols. 2d ed. Original Arabic text, 1396. Princeton, NJ: Princeton University Press, 1967.

Lenski, Gerhard. *Ecological-Evolutionary Theory.* Boulder, CO: Paradigm Publishers, 2005.

Leon-Portilla, Miguel. *Native Meso-American Spirituality.* New York: Paulist Press, 1980.

McKibben, Bill. *The End of Nature.* New York: Doubleday, 1989.

McNeill, J. R. *Something New Under the Sun: An Environmental History of the Twentieth-Century.* New York: Norton, 2000.

McNeill, John, and William Hardy McNeill. *The Human Web: A Birds-Eye View of World History.* New York: Norton, 2003.

Moore, Kathleen Dean. *Holdfast: At Home in the Natural World.* New York: Lyons Press, 1999.

Needham, Joseph. *Science and Civilisation in China.* 6 vols. Cambridge: Cambridge University Press, 1954–1988.

Norgaard, Richard B. *Development Betrayed: The End of Progress and a Coevolutionary Revisioning of the Future.* London and New York: Routledge, 1994.

Novacek, Michael. *Terra: Our 100-Million-Year-Old Ecosystem and the Threats That Now Put It at Risk.* New York: Farrar, Straus and Giroux, 2007.

Orr, David. *Down to the Wire: Confronting Climate Collapse.* New York: Oxford University Press, 2009.

Palumbo, Stephen. *The Evolution Explosion: How Humans Cause Rapid Evolutionary Change.* New York: W. W. Norton, 2001.

Rockefeller, Steven, and John C. Elder. *Spirit and Nature: Why the Environment Is a Religious Issue.* Boston: Beacon Press, 1992.

Roszak, Theodore. *The Voice of the Earth.* Grand Rapids, MI: Phanes Press, 1992.

Sanders, Scott Russell. *Hunting for Hope.* Boston: Beacon Press, 1998.

Schellnhuber, Hans Joachim, Paul J. Crutzen, William C. Clark, Martin Claussen, and Hermann Held, eds. *Earth System Analysis for Sustainability.* Cambridge, MA: MIT Press, 2004.

Sullivan, Lawrence Eugene. *Icanchu's Drum: An Orientation to Meaning in South American Religions.* New York: Macmillan, 1988.

Tarnas, Richard. *The Passion of the Western Mind: Understanding the Ideas That Have Shaped Our World View.* New York: Harmony, 1991.

Tattersall, Ian. *Paleontology: A Brief History of Life.* West Conshohocken, PA: Templeton Press, 2010.

Tirosh-Samuelson, Hava, ed. *Judaism and Ecology: Created World and Revealed Word.* Religions of the World and Ecology Series, edited by Mary Evelyn Tucker and John Grim. Cambridge, MA: Harvard Center for the Study of World Religions, 2002.

Tucker, Mary Evelyn. *Philosophy of Qi.* New York: Columbia University Press, 2007.

Tucker, Mary Evelyn, and John Berthrong, eds. *Confucianism and Ecology: The Interrelation of Heaven, Earth, and Humans.* Religions of the World and Ecology Series, edited by Mary Evelyn Tucker and John Grim. Cambridge, MA: Harvard Center for the Study of World Religions, 1998.

Tucker, Mary Evelyn, and Duncan Ryuken Williams, eds. *Buddhism and Ecology: The Interconnection of Dharma and Deeds.* Religions of the World and Ecology Series, edited by Mary Evelyn Tucker and John Grim. Cambridge, MA: Harvard Center for the Study of World Religions, 1997.

Wright, Robert. *Nonzero: The Logic of Human Destiny.* New York: Vintage, 2001.

T E N   Rethinking Matter and Time

Baaquie, Belal, and Frederick Willeboordse. *Exploring Integrated Science.* Boca Raton: CRC Press, 2010.

Baggott, Jim. *Beyond Measure: Modern Physics, Philosophy and the Meaning of Quantum Theory.* Oxford: Oxford University Press, 2004.

Barlow, Connie, ed. *Evolution Extended: Biological Debates on the Meaning of Life.* Cambridge, MA: MIT Press, 1994.

Barrow, John, Simon Conway Morris, Stephen Freeland, and Charles Harper, eds. *Fitness of the Cosmos for Life: Biochemistry and Fine-Tuning.* Cambridge: Cambridge University Press, 2007.

Beinhocker, Eric. *The Origin of Wealth: Evolution, Complexity, and the Radical Remaking of Economics.* Boston: Harvard Business School Press, 2006.

Bennett, Jane. *Vibrant Matter: A Political Ecology of Things.* Durham, NC: Duke University Press, 2010.

Bergson, Henri. *Creative Evolution.* Westport, CT: Greenwood Press, 1975.

Bird, Richard J. *Chaos and Life: Complexity and Order in Evolution and Thought.* New York: Columbia University Press, 2003.

Bokulich, Alisa. *Reexamining the Quantum: Classical Relation.* Cambridge: Cambridge University Press, 2008.

Bruteau, Beatrice. *God's Ecstasy: The Creation of a Self-Creating World.* New York: Crossroad, 1997.

Capra, Fritjof. *The Hidden Connections: Integrating the Biological, Cognitive, and Social Dimensions of Life into a Science of Sustainability.* New York: Doubleday, 2002.

Carson, Rachel. *Silent Spring.* Anniversary edition. New York: Mariner Books, 2002 (originally published in 1962).

Clayton, Philip. *Mind and Emergence: From Quantum to Consciousness.* New York: Oxford University Press, 2006.

Davies, Paul, and Niels Gregersen, eds. *Information and the Nature of Reality.* Cambridge: Cambridge University Press, 2010.

Dick, Steven, and Mark Lupisella, eds. *Cosmos & Culture: Cultural Evolution in a Cosmic Context.* Washington, DC: NASA SP-4802, http://history.nasa.gov/SP-4802.pdf, 2010.

Ehrlich, Gretel. *Solace of Open Spaces.* New York: Penguin, 1985.

Eiseley, Loren. *The Immense Journey.* New York: Vintage, 1956.

Holland, John. *Emergence.* Reading, MA: Addison-Wesley, 1998.

Holmes, Barbara A. *Race and the Cosmos: An Invitation to View the World Differently.* Harrisburg, PA: Trinity Press International, 2002.

Impey, Chris. *The Living Cosmos: Our Search for Life in the Universe.* New York: Random House, 2007.

Jantsch, Erich. *The Self-Organizing Universe: Scientific and Human Implications of the Emerging Paradigm of Evolution.* New York: Pergamon Press, 1980.

Jencks, Charles. *The Garden of Cosmic Speculation.* London: Frances Lincoln, 2003.

Johnson, Steven. *Emergence: The Connected Lives of Ants, Brains, Cities, and Software.* New York: Scribner, 2002.

Kauffman, Stuart. *At Home in the Universe.* New York: Oxford University Press, 1995.

Kelso, J. A. Scott, and David Engstrom. *The Complementary Nature.* Cambridge, MA: MIT Press, 2006.

Leopold, Aldo. *Sand County Almanac.* New York: Oxford University Press, 1966 (first published in 1949).

Lockwood, Michael. *The Labyrinth of Time: Introducing the Universe.* Cambridge: Cambridge University Press, 2007.

Lopez, Barry. *River Notes: The Dance of the Herons.* New York: Avon Books, 1979.

Mainzer, Klaus. *Thinking in Complexity: The Computational Dynamics of Matter, Mind, and Mankind.* Berlin: Springer, 2007.

Miller, James Grier. *Living Systems.* New York: McGraw-Hill, 1978.

Mitchell, Melanie. *Complexity: A Guided Tour.* Oxford: Oxford University Press, 2009.

Morowitz, Harold. *The Emergence of Everything: How the Universe Became Complex.* New York: Oxford University Press, 2004.

Nadeau, Robert, and Menas Kafatos. *The Non-local Universe: The New Physics and Matters of the Mind.* New York: Oxford University Press, 2001.

Nunez, Paul. *Brain, Mind, and the Structure of Reality.* Oxford: Oxford University Press, 2010.

Oliver, Mary. *New and Selected Poems.* Boston: Beacon Press, 1992.

Prigogine, Ilya. *From Being to Becoming: Time and Complexity in the Physical Sciences.* San Francisco: Freeman, 1980.

Prigogine, Ilya, and Isabelle Stengers. *Order Out of Chaos: Man's New Dialogue with Nature.* New York: Bantam Books, 1984.

Rogers, Patiann. *Fire-keepers: New and Selected Poems.* Minneapolis, MN: Milkweed, 1994.

Scott, Alwyn. *The Nonlinear Universe.* Berlin: Springer, 2007.

Skrbina, David, ed. *Mind That Abides: Panpsychism in the New Millennium.* Amsterdam: John Benjamins, 2009.

Snyder, Gary. *Back on the Fire.* Berkeley: Shoemaker and Hoard, 2007.

Swimme, Brian, and Thomas Berry. *The Universe Story.* San Francisco: HarperSanFrancisco, 1992.

Teilhard de Chardin, Pierre. *The Human Phenomenon.* Portland, OR: Sussex Academic Press, 2003.

Thompson, Evan. *Mind in Life: Biology, Phenomenology, and the Sciences of Mind.* Cambridge, MA: Harvard University Press, 2007.

Toolan, David. *At Home in the Cosmos.* Maryknoll, NY: Orbis Books, 2001.

Toulmin, Stephen Edelston. *The Return to Cosmology: Postmodern Science and the Theology of Nature.* Berkeley: University of California Press, 1982.

Toulmin, Stephen, and June Goodfield. *The Discovery of Time.* Chicago: University of Chicago Press, 1977.

Wheeler, Wendy. *The Whole Creature: Complexity, Biosemiotics and the Evolution of Culture.* London: Lawrence and Wishart, 2006.

Whitehead, Alfred North. *Process and Reality: An Essay in Cosmology.* New York: Free Press, 1929.

Whitfield, John. *In the Beat of a Heart: Life, Energy, and the Unity of Nature.* Washington, DC: Joseph Henry Press, 2006.

Williams, Terry Tempest. *Refuge.* New York: Vintage Books, 1991.

ELEVEN   Emerging Earth Community

Benyus, Janine. *Biomimicry.* New York: William Morrow, 1997.

Berkes, Firket. *Sacred Ecology.* 2d ed. New York: Routledge, 2008.

Berry, Thomas. *The Dream of the Earth.* San Francisco: Sierra Club Books, 1988.

——. *The Sacred Universe.* Ed. Mary Evelyn Tucker. New York: Columbia University Press, 2009.

Birkeland, Janis. *Positive Development: From Vicious Cycles Through Built Environmental Design.* London: Earthscan, 2008.

Bolen, Jean Shinoda. *Urgent Message from Mother: Gather the Women, Save the World.* Boston: Conari Press, 2005.

Bronk, Richard. *The Romantic Economist.* Cambridge: Cambridge University Press, 2009.

Brown, Brian. *Religion, Law, and the Land: Native Americans and the Judicial Interpenetration of Sacred Land.* Westport, CT: Greenwood Press, 1999.

Brown, Cynthia. *Big History: From the Big Bang to the Present.* New York: The New Press, 2007.

Brown, Lester R. *Plan B 4.0: Mobilizing to Save Civilization.* New York and London: W. W. Norton, 2009.

Brown, Peter, and Geoffrey Garver. *Right Relationship: Building a Whole Earth Economy.* San Francisco: Berrett-Koehler, 2009.

Carson, Rachel. *Silent Spring.* Twenty-fifth anniversary edition. Boston: Houghton Mifflin, 1987.

Corcoran, Peter, editor-in-chief. *Toward a Sustainable World: The Earth Charter in Action.* Amsterdam: KIT Publishers, 2005.

Crist, Eileen, and H. Bruce Rinker, eds. *Gaia in Turmoil: Climate Change, Biodepletion, and Earth Ethics in an Age of Crisis.* Cambridge, MA: MIT Press, 2010.

Csikszentmihalyi, Mihaly. *The Evolving Self: A Psychology for the Third Millennium.* New York: HarperCollins, 1993.

Cullinan, Cormac. *Wild Law: A Manifesto for Earth Justice.* Devon, U.K.: Green Books, 2003.

Dalton, Ann Marie, and Henry Simmons. *Ecotheology and the Practice of Hope.* Albany, NY: SUNY Press, 2010.

Daly, Herman E. *Beyond Growth: The Economics of Sustainable Development.* Boston: Beacon Press, 1997.

Dellinger, Drew. *Love Letter to the Milky Way: A Book of Poems.* Mill Valley, CA: Planetize the Movement Press, 2002.

Diamond, Irene, and Gloria Femen Orenstein. *Reweaving the World: The Emergence of Ecofeminism.* San Francisco: Sierra Club Books, 1990.

Downton, Paul. *Ecopolis: Architecture and Cities for a Changing Climate.* Dordrecht: Springer, 2009.

Eaton, Heather, and Lois Ann Lorentzen, eds. *Ecofeminism and Globalization.* New York: Rowan and Littlefield, 2003.

Ehrlich, Paul, and Anne Ehrlich. *One with Ninevah: Politics Consumption and the Human Future.* Washington, DC: Island Press, 2004.

Fox, Matthew. *The A.W.E. Project: Reinventing Education, Reinventing the Human.* Kelowna, B.C.: CopperHouse, 2006.

Frank, Adam. *The Constant Fire: Beyond the Science vs. Religion Debate.* Berkeley: University of California Press, 2009.

Genet, Cheryl, Russell Genet, Brian Swimme, Linda Palmer, and Linda Gibler. *The Evolutionary Epic: Science's Story and Humanity's Response.* Santa Margarita, CA: Collins Foundation Press, 2009.

Goerner, Sally, Robert G. Dyck, and Dorothy Lagerroos. *The New Science of Sustainability: Building a Foundation for Great Change.* Chapel Hill, NC: Triangle Center for Complex Systems, 2008.

Goodall, Jane. *Reason for Hope: A Spiritual Journey.* New York: Warner Books, 1999.

Goodwin, Brian. *Nature's Due: Healing Our Fragmented Culture.* Edinburgh, U.K.: Floris Books, 2007.

Gore, Al. *Our Choice: A Plan to Solve the Climate Crisis.* Emmaus, PA: Rodale Press, 2009.

Gregg, Gary. *The Middle East: A Cultural Psychology.* New York: Oxford University Press, 2005.

Griffin, Susan. *Women and Nature: The Roaring Inside Her.* New York: Harper and Row, 1978.

Haberman, David. *River of Love in an Age of Pollution.* Berkeley, CA: University of California Press, 2006.

Hamilton, Clive. *Growth Fetish.* London: Pluto Press, 2003.

Hathaway, Mark, and Leonardo Boff. *The Tao of Liberation: Exploring the Ecology of Transformation.* Maryknoll, NY: Orbis Books, 2009.

Haught, John. *God After Darwin: A Theology of Evolution.* Boulder, CO: Westview Press, 2001.

Hawken, Paul. *Blessed Unrest.* New York: Viking, 2007.

Henderson, Hazel. *Paradigms in Progress: Life Beyond Economics.* Indianapolis: Knowledge Systems Incorporated, 1991.

Holthaus, Gary. *From the Table to the Farm: What All Americans Need to Know About Agriculture.* Lexington: University of Kentucky, 2007.

Hyams, Edward. *Soil and Civilization.* New York: State Mutual Books, 1980.

Jackson, Wes. *Becoming Native to This Place.* Lexington: University of Kentucky, 1994.

Jenkins, Willis, ed. *The Spirit of Sustainability.* Vol. 1 of *Berkshire Encyclopedia of Sustainability.* Great Barrington, MA: Berkshire Publishing, 2009.

Jordon, William R. *The Sunflower Forest: Ecological Restoration and the New Communion with Nature.* Berkeley, CA: University of California Press, 2003.

Kaufman, Gordon. *In Face of Mystery: A Constructive Theology.* Cambridge, MA: Harvard University Press, 1995.

Keller, Catherine. *On the Mystery: Discerning God in Process.* Minneapolis, MN: Fortress, 2008.

Kelly, Sean M. *Coming Home: The Birth and Transformation of the Planetary Era.* Great Barrington: Lindisfarne Books, 2010.

Korten, David. *The Great Turning: From Empire to Earth Community.* San Francisco: Berrett-Koehler, 2006.

Leopold, Aldo. *A Sand County Almanac.* New York: Oxford University Press, 1949.

Levin, Simon, ed. *Games, Groups, and the Global Good.* Berlin: Springer, 2009.

Litfin, Karen. "Towards an Integral Perspective on World Politics: Secularism, Sovereignty and the Challenge of Global Ecology." *Millennium: Journal of International Studies* 32/1 (2003): 29–56.

Louv, Richard. *Last Child in the Woods: Saving Our Children from Nature-Deficit Disorder.* Chapel Hill, NC: Algonquin Books, 2005.

Maathai, Wangari. *The Green Belt Movement.* New York: Lantern Books, 2003.

McDermott, Robert A., and V. S. Naravane, eds. *The Spirit of Modern India: Writings in Philosophy, Religion, and Culture.* Herndon, VA: Lindisfarne Books, 2010.

McKibben, Bill. *Eaarth: A Guide to Living on a Fundamentally Altered Planet.* New York: Times Books, 2010.

Meadows, Donella, Jorgen Randers, and Dennis L. Meadows. *Limits to Growth: The 30-Year Update.* White River Junction, VT: Chelsea Green, 2004.

Merchant, Carolyn. *The Death of Nature: Women, Ecology and the Scientific Revolution.* San Francisco: Harper and Row, 1981.

Miller, James, ed. *The Epic of Evolution: Science and Religion in Dialogue.* AAAS Conference at the Field Museum. Upper Saddle River, NJ: Prentice Hall, 2004.

Moore, Kathleen Dean, and Michael P. Nelson, eds. *Moral Ground: Ethical Action for a Planet in Peril.* San Antonio: Trinity University Press, 2010.

Nash, Roderick. *The Rights of Nature: A History of Environmental Ethics.* Madison: University of Wisconsin Press, 1989.

Nelson, Melissa K., ed. *Original Instructions: Indigenous Teachings for a Sustainable Future.* Rochester, VT: Bear and Company/Inner Traditions, 2008.

Orr, David. *Ecological Literacy.* Albany, NY: SUNY Press, 1992.

Pavel, Paloma, ed. *Breakthrough Communities: Sustainability and Justice in the Next American Metropolis.* Cambridge, MA: MIT Press, 2009.

Rasmussen, Larry. *Earth Community, Earth Ethics.* Maryknoll, NY: Orbis Books, 1996.

Raymo, Chet. *Skeptics and True Believers: The Exhilarating Connection Between Science and Religion.* New York: Walker and Co., 1998.

Register, Richard. *Ecocities: Building Cities in Balance with Nature.* Berkeley, CA: Beverly Hills Books, 2002.

Rifkin, Jeremy. *The Empathic Civilization: The Race to Global Consciousness in a World in Crisis.* New York: Tarcher/Penguin, 2010.

Ritter, Dale F., R. Craig Kochel, and Jerry R. Miller. *Process Geomorphology.* 5th ed. Long Grove, IL: Waveland Press, 2011.

Rue, Loyal. *Everybody's Story: Wising Up to the Epic of Evolution.* Albany: State University of New York, 2000.

Sachs, Jeffrey. *Common Wealth: Economics for a Crowded Planet.* New York: Penguin Press, 2008.

Scheffer, Marten. *Critical Transitions in Nature and Society.* Princeton: Princeton University Press, 2009.

Schell, Jonathan. *The Seventh Decode: The New Shape of Nuclear Danger.* New York: Metropolitan Books, 2008.

Schmitz, Oswald. *Ecology and Ecosystem Conservation.* Washington, DC: Island Press, 2007.

Schor, Juliet. *Plenitude: The New Economics of True Wealth.* New York: Penguin Press, 2010.

Selin, Helaine. *Nature Across Cultures: Views of Nature and the Environment in Non-Western Cultures.* Berlin: Springer, 2003.

Shiva, Vandana. *Soil Not Oil: Environmental Justice in an Age of Climate Crisis.* Cambridge, MA: South End Press, 2008.

Speth, James Gustave. *The Bridge at the End of the World: Capitalism, the Environment, and Crossing from Crisis to Sustainability.* New Haven: Yale University Press, 2008.

Spier, Fred. *Big History and the Future of Humanity.* Malden, MA: Wiley-Blackwell, 2010.

Spretnak, Charlene. *Resurgence of the Real.* New York: Routledge, 1999.

Thomashow, Mitchell. *Bringing the Biosphere Home: Learning to Perceive Global Environmental Change.* Cambridge, MA: MIT Press, 2003.

Tu Weiming. *Commonality and Centrality: An Essay on Confucian Religiousness.* Albany, NY: SUNY Press, 1989.

Tucker, Mary Evelyn. *Worldly Wonder: Religions Enter Their Ecological Phase.* LaSalle, IL: Open Court, 2004.

Tucker, Mary Evelyn, and John Grim, eds. *Worldviews and Ecology: Religion, Philosophy and the Environment.* Maryknoll, NY: Orbis Books, 2008. First published 1994.

Waltner-Toews, David, James J. Kay, and Nina-Marie E. Lister, eds. *The Ecosystem Approach: Complexity, Uncertainty, and Management for Sustainability.* New York: Columbia University Press, 2008.

Warren, Julianne Lutz. *Aldo Leopold's Odyssey.* Washington, DC: Island Press, 2006.

Wilson, Edward O. *Biophilia: The Human Bond with Other Species.* Cambridge, MA: Harvard University Press, 1984.

——. *The Creation: An Appeal to Save Life on Earth.* New York: Norton, 2007.

Worster, Donald. *Nature's Economy: The Roots of Ecology.* Garden City, NY: Anchor Press/Doubleday, 1977.

INDEX